T0211693

SpringerBriefs in Applied Sciences and Technology

More information about this series at http://www.springer.com/series/8884

Christina W.Y. Wong · Kee-hung Lai
Y.H. Venus Lun · T.C.E. Cheng

Environmental Management

The Supply Chain Perspective

 Springer

Christina W.Y. Wong
Institute of Textiles and Clothing
The Hong Kong Polytechnic University
Hong Kong
China

Kee-hung Lai
Department of Logistics and Maritime
 Studies
The Hong Kong Polytechnic University
Hong Kong
China

Y.H. Venus Lun
Department of Logistics and Maritime
 Studies
The Hong Kong Polytechnic University
Hong Kong
China

T.C.E. Cheng
Department of Logistics and Maritime
 Studies
The Hong Kong Polytechnic University
Hong Kong
China

ISSN 2191-530X   ISSN 2191-5318 (electronic)
SpringerBriefs in Applied Sciences and Technology
ISBN 978-3-319-23680-3   ISBN 978-3-319-23681-0 (eBook)
DOI 10.1007/978-3-319-23681-0

Library of Congress Control Number: 2015948729

Springer Cham Heidelberg New York Dordrecht London

Printed on acid-free paper

Springer International Publishing AG Switzerland is part of Springer Science+Business Media
(www.springer.com)

Preface

The urgency of reducing the environmental impact of human activities and conserving natural resources has been recognized and considered one of the key agenda items of many organizations. While firms are investing in environmental management, collaborative efforts across supply chain partners are needed to achieve desirable results. This book presents theory-driven discussions on the link between environmental management and business performance in the context of supply chain management. This book provides knowledge from the research findings and real-life cases that helps managers and students develop plans to implement environmental management practices jointly with supply chain partners. This book serves as a good reference to practitioners and students to understand how to manage supply chains to improve business as well as environmental performance. This book consists of five chapters covering topics on environmental management, environmental management practices with supply chain efforts, collaborative environmental management, organizational capabilities in environmental management, and closed-loop supply chains.

By introducing the concept of environmental management in a supply chain management context, organizations consider their environmental impact in the various stages in a supply chain, ranging from product design, cleaner production, distribution, to end-of-life product treatments in order to reduce waste, replace use of harmful materials, recycle, and reuse. To achieve these ends, the involvement of suppliers and customers is critical as they are taking part in a supply chain. The involvement of suppliers and customers requires collaborative efforts, including environmental information sharing, operational adaptation across processes, integration, and understanding of market preferences.

In addition to supply chain collaboration for environmental management, this book also discusses the importance of environmental disclosure that publicizes information related to organizational efforts in deterring environmental impact, while communicating with supply chain partners on their positioning in environmental protection. As green advertising is an approach of environmental disclosure, this book also discusses the research findings of the performance impact of green advertising. While organizational capabilities also play important role in environmental management, the importance of intra-organizational integration, environmental innovativeness, and environmental adaptability to business and environmental performance is discussed with case examples.

The four co-authors of this book have substantial research experience in the area of environmental management and supply chain management. The discussion of the subjects in this book was their experience and research findings.

Contents

About the Authors

Christina W.Y. Wong is an associate professor, specializing in supply chain management and environmental management, in the Institute of Textiles and Clothing, The Hong Kong Polytechnic University. She has co-authored over 60 papers in such journals as *Nature, Production and Operations Management, Journal of Operations Management, Journal of Management Information Systems*, and others.

Kee-hung Lai is an associate professor in the Department of Logistics and Maritime Studies, The Hong Kong Polytechnic University. He has co-authored five books and over 110 papers in such scholarly journals as *California Management Review, Journal of Operations Management, Production & Operations Management*, and others. He is the winner of the first annual ProSPER.Net-Scopus Young Scientist Award on sustainable development in the business category in 2009.

Y.H. Venus Lun is director of Shipping Research Center and assistant professor in the Department of Logistics and Maritime Studies at The Hong Kong Polytechnic University. She has co-authored six books published by Inderscience, McGraw-Hill, Springer, and VDM. She has also published more than 40 papers in scholarly journals.

T.C.E. Cheng is dean of the Faculty of Business, Fung Yiu King—Wing Hang Bank Professor in Business Administration, and chair professor of Management at The Hong Kong Polytechnic University. He has published over 600 papers in such journals as *California Management Review, Journal of Operations Management, Management Science, MIS Quarterly, Operations Research, Organization Science*, and *Production and Operations Management*, and co-authored 11 books published by Chapman and Hall, McGraw-Hill, and Springer.

Chapter 1
Environmental Management

Abstract This chapter introduces the basic concepts of environmental management, and extends the discussion to the selected topics in green supply chain management, environmental disclosure, and environmental reputation. Environmental management is a management practice of reducing environmental impacts of organizations, which includes practices such as eco-design, waste reduction, recycling, reuse, adoption of cleaner technologies, and green logistics. The diffusion of environmental management systems across different organizational functions plays a key role in driving environmental management practices. Organization that has higher capability of environmental management can achieve better financial performance. Successful environmental management also requires vertical coordination and cooperation between suppliers and customers in supply chain. A positive environmental reputation is a valuable intangible asset of organizations that reflects successful fulfillment of stakeholders' expectations. Although green advertising may be detrimental to organization by bringing about costs, there are also beneficial effects of publicizing information related organizational efforts on environmental protection. Overall, environmental management helps organizations to sustain economic viability without depleting environmental resources, and have been prioritized in the business world including big brands in electronics, consumer goods, transportation, and other industries.

1.1 Concepts of Environmental Management

1.1.1 Definition of Environmental Management

Environmental management is a system of functions that are used to develop, implement, and monitor organizational environmental strategies to achieve environmental objectives [1]. Corporate environmental objectives are concerned with the organizational goals that were set to prevent, abate, minimize, and remedy

© The Author(s) 2016
C.W.Y. Wong et al., *Environmental Management*,
SpringerBriefs in Applied Sciences and Technology,
DOI 10.1007/978-3-319-23681-0_1

environmental damages caused by organizational operations. In most cases, environmental management is related to taking steps and promoting behaviors that reduce environmental impact in conducting business activities.

The traditional focus of managing environmental issues was to comply with environmental regulations. The main theme of corporate environmental management included putting in quest for permits, monitoring the pollutant emissions, reporting to government to meet regulations, and avoiding fines [2]. However, this traditional relationship between firms and government has been shifting over the decades. More voluntary and flexible regulatory policies (e.g. cap-and-trade) have been introduced to allow for market and economic incentive to manage pollution. Nowadays firms are engaging in environmental management for various reasons, including caring for the natural resources conservation, meeting the needs of environmentally-aware consumers, and saving economic expenses [1, 3].

Many businesses have a strong commitment to the excellence in environmental management. Take Apple, the world-leading innovator in computing technology, for example. Apple strives to be leading-edge in both product quality improvement and environmental performance. Apple introduced its new 17-inch MacBook Pro that is expected to last for five years instead of two years, while the battery life will last longer than the typical cycle for laptops. Apple is also successful in eliminating many harmful toxins in its products. It managed to remove mercury from the cold cathode fluorescent lamp (CFL) backlights by switching to the light-emitting diodes (LED) backlights, and removed arsenic contained in the glass of conventional LED displays. It also reduces the size of packaging boxes by 40 % to mitigate costs as well as wastes. The enclosure is made of recyclable aluminum while the screen is made of recyclable glass. These modest improvements amount to substantial environmental improvements of Apple [1].

1.1.2 Environmental Management Practices

Facing a variety of environmental pressures, firms recognize the importance of environmental management. In the literature, environmental management practices are concerned with different activities in reducing damages caused to the environment, including such environmental initiatives as eco-design of products [3–5], recycling and waste reduction [5–7], adoption of environmental technologies [8, 9], green logistics and transportation [10–12].

Eco-design can be referred as preventive measure of environmental management practice [4]. For example, Philips, a Dutch consumer technology company, has been methodically reducing use of toxic chemicals in its products and minimizing the environmental impact of its operations through research and product design. It strives for its green products to account for 55 % of revenue by 2015. One relatively recent addition to its lineup is the Pacific LED Green Parking system, which includes integrated luminaires, wireless controls, and presence detection applications. Philips pitches the platform as both a safety play and a cost-effective replacement for traditional fluorescent lamps, reducing consumption of electricity when its products are in use [13].

Recycling is an environmental management practice that captures the residual materials from used products and reuses the materials to produce new products in order to reduce the consumption of raw materials, prevent waste of potentially useful materials, and mitigate pollution by lowering the conventional waste disposal [11]. Recycling allows firms to disassemble components and parts for capturing residual values of returned products and reduce consumption of new inputs by using reusable parts captured from returned products [14]. As a leading producer of specialty polymeric compounds, Spartech Corporation of St. Louis has integrated recycling into its business practices since it began manufacturing in 1960. By reprocessing plastics into new products and sweeping up plastic pellets regularly, Spartech salvaged more than two million pounds of waste that would otherwise end up in landfills each year, thereby saving millions of dollars. Spartech also saved more than US$211,000 from landfill and packaging costs by recycling corrugated cardboard and wooden pallets [15].

Environmental technologies are defined as equipment, methods and procedures, product designs, and product delivery mechanisms that conserve energy and natural resources, minimize environmental load of human activities, and protect the natural environment. They are evolving from both as a set of technologies, equipment, and operating procedures and as a management orientation towards technology choice and design of industrial systems [7–9]. Take the adoption of textile dyeing techniques as an example. According to the World Bank, about 72 toxic chemicals in water are generated from textile dyeing, 30 of which cannot be removed during the process of wastewater treatment by conventional techniques. After eight-year exploration, Nike developed a waterless dyeing technology that makes use of recycled carbon dioxide to color synthetic textiles. This technology is able to eliminate billions of gallons of polluted wastewater discharged into waterways near manufacturing plants of Asia where much of the world's textile dyeing occurs. Nike's adoption of waterless dyeing process helps to alleviate the global concerns over water scarcity and waterways pollution [16].

Green logistics and transportation are environmental management practices that are deployed by firms to distribute goods sustainably in order to reduce waste and conserve resources in performing logistics activities [7, 17]. JohnsonDiversey, a company that makes cleaning and food safety products, has put in place a series of projects to cut logistics-related greenhouse gas emissions by a quarter, while reduce lead times on product deliveries and boost service levels. JohnsonDiversey ships chemical and paper orders to customers on the same truck to fully utilize truck space and to reduce emissions and fuel consumption. Truck movement between JohnsonDiversey's manufacturing plant and warehouse was drastically reduced with the installation of a lift and conveyor system that connects the two facilities. Forklift operations are eliminated in the facilities, leading to less energy consumption. As a result, JohnsonDiversey is able to reduce emissions by 457 metric tons and save 42,267 gallons of fuel every year [18].

1.1.3 Environmental Management Diffusion

Environmental management system (EMS) is a set of practices that help a company reduce its environmental impacts while increase its operating efficiency. More formally, EMS is a system and database which integrates processes for training of personnel, monitoring, summarizing, and reporting of specialized environmental performance information to internal and external stakeholders of a company [19]. An EMS encourages a company to constantly improve its environmental performance. The most widely used framework for an EMS is the one developed by the International Organization for Standardization (ISO) for the ISO 14,000 family which is concerned with various aspects of environmental management. The only standard that may be registered or certificated is the ISO 14001 standard which is established in 1996. The ISO 14001 standard defines five stages of an EMS, including commitment and policy, planning, implementation, checking & monitoring, and review [2].

Diffusion of EMS is referred as the extent to which the requirements of ISO 14001 have impacted on different functions of the organizations which results in changes of the practices performed by those functions [20]. It is found that the diffusion of EMS within organizations has a positive effect on green practices in the form of green products, green processes, and green supply chain management [20]. Different approaches in diffusing the ISO 14001 standard will affect the outcomes. Firms may adopt decoupled integration took the minimalist approach in meeting ISO 14001 requirements, thereby bringing little change in the existing practices. But for firms which took mobilized and proactive integration, the involvement of employees and managers in implementing the ISO 14001 standards will help to significantly change the environmental management practices [21]. From a cross-functional perspective, ISO 14001-based EMS generates diffusion on five key organizational functions, namely production, procurement, sales, logistics, and R&D. In addition, diffusion imbalance can harm green product and green supply chain management [20].

1.1.4 Environmental Management Capability

Environmental management capability is referred to as organizational ability to operate business in an environmentally-friendly manner while pursuing financial gains [22]. Such capability is often characterized with the adoption of environmental management systems such as ISO 14000, evaluation of the suppliers' environmental performance, and development of an environmental policy to reduce environmental footprints [23]. Recent studies show that environmental management capability is valuable to firms that rely heavily on the supply network to develop complicated electronics products, provide value-added services, implement complex business processes, and meet higher customer expectations [24–27]. Firms that have knowledge and capabilities to anticipate, adapt, affect, and respond to the environmental policy can build salient comparative advantages over their counterparts [2].

Panasonic wields enormous influence globally in the areas of forward-moving technologies of energy-efficient lighting, refrigeration, heating, air-conditioning, and renewable energy production. Panasonic keep tracks of the energy consumption of its facilities using ISO 14000 systems. In particular, its U.S. operating unit is able to reduce its carbon footprint by at least 50 % compared with its old facility. The operations process is engineered specifically to live the commitment to the Panasonic's principles of combining innovation, collaboration, and sustainability. This combination has led Panasonic's Eco Solutions group to assume a relatively unique role: end-to-end management of renewable energy projects for commercial, government and industrial customers. Panasonic's customers also urges Panasonic to engineer custom eco-solutions by partnering with customers to achieve results that leverage its high-end products and onsite services [28].

1.1.5 Summary

Environmental management is a management practice that organizations undertake to regulate and protect the natural environment. It promotes organizational efforts to reduce impact on the environment, while saving and utilizing natural resources. Specifically, environmental management includes such practices as eco-design, recycling and waste reduction, adoption of environmental technologies, and green logistics and transportation, and so forth. EMS sets up a framework to help companies achieve their environmental goals through consistent controlling its production operations. In particular, ISO 14001 standard provides practical tools for organizations to control their environmental impact and continuously improve their environmental performance. The balanced diffusion of EMS across different organizational functions is an important factor in driving environmental management practices. Firms with capability of environmental management are able to achieve financial gains with minimum compromise to the environment.

1.2 Environmental Management and Supply Chain Management

1.2.1 Corporate Environmental Management

Corporate environmental management is referred as the organizational efforts and practices to reduce adverse environmental impact through product and process stewardship with an emphasis on reducing liability and costs [29]. Corporate environmental management helps reduce legislative non-compliance, improve staff environmental awareness, ensure continual environmental improvement, and raise financial benefits by resource saving and cost reduction. From the perspective of public relations, corporate environmental management can enhance organizational reputation, and establish favorable customer relationships [30, 31].

The Boeing Company is an American multinational corporation that designs, manufactures, and sells fixed-wing aircraft, rotorcraft, rockets, and satellites. It also provides leasing and product support services. Since January 2014, Boeing expands the use of sustainable aviation biofuel by seeking approval of U.S. Federal Aviation Administration (FAA) for its aircraft to fly on green diesel. Green diesel is a fuel derived from feedstock, such as non-food plant oils, waste cooking oil and animal fats, and algae. This fuel is chemically similar to sustainable aviation biofuels that emits at least half the lifecycle carbon dioxide as traditional petroleum-based jet fuel [32].

Best Buy, an American consumer electronics corporation, has put a lot of efforts in investing in electronics recycling space, setting carbon footprint reduction goals, implementing environmental management systems ISO 14001, and participating in federal environmental policy legislation. With electronics recycling, for example, Best Buy realized most customers coming into their stores are replacing old products. It therefore set up recycling kiosks in the front of all of its stores. Best Buy accepts electronics at no cost, regardless of where the items originally were purchased. As a result, Best Buy recycled more than 21 million units in 2012. In terms of their carbon footprint reduction work, Best Buy is one of the original 14 members to join the White House Better Buildings Challenge, which challenges companies in the private sector to reduce energy consumption as a way to address climate change. In 2010, Best Buy set a goal to reduce absolute emissions by 20 % by 2020 and reached 75 % of that goal by the end of 2012 [33].

1.2.2   Green Supply Chain Management

Supply chain management is referred to as the management of design, planning, execution, control, and monitoring of supply chain activities with the objective of creating net value, building a competitive infrastructure, leveraging worldwide logistics, synchronizing supply with demand [34]. Supply chain management handles the movement and storage of raw materials, work-in-process inventory, and finished goods from the point of origin to the point of consumption. Interconnected or interlinked networks, channels and node businesses are involved in the provision of products and services to end customers in a supply chain [35]. The global supply chain with various partners exposes a firm to the increased risk, cost, inefficiency, waste, and environmental impact.

Green supply chain management (GSCM) is concerned with inter-organizational efforts in managing supply chain activities to lower the negative environmental impact from purchasing of materials, manufacturing, to distribution of finished products [36]. From a system-wide perspective, GSCM not only represents the relationships between customers and their suppliers, but also includes multiple stages of the supply chain and the outbound logistics [2]. Due to the increasing pressures for corporate environmental sustainability, GSCM has received significant attention among businesses. Following are three exemplars of GSCM implementation:

- Drawing upon its IT expertise, IBM cooperates with its suppliers to create a new consulting service to help its clients to improve operations efficiency and reduce environmental impacts of their supply chains. The Sustainable Supplier Information Management Consulting service help firms collect supply chain data, such as energy use, labor practices, and greenhouse gas emissions. With IBM's service, Wal-Mart is able to collect environmental data from its top-tier suppliers. The new service also helps firms develop the capability to track part numbers, audit supplier environmental performance, and vet new suppliers. The adoption of this service is expected to yield efficiency and reduce cost of at least 8 % [37].
- Kaiser Permanente is the largest integrated managed care consortium based in the US. Acknowledging that sustainability data is factored into purchasing decisions, Kaiser requires its suppliers to provide environmental data on all medical equipment and products. Kaiser develops its Sustainability Scorecard system that incorporates the environmental impacts of producing, obtaining, using and disposing of medical products as integral to purchasing decisions. With the scorecard, Kaiser tracks the environmental impacts of its products. [38].
- Marks & Spencer (M&S) is a major British multinational retailer specializing in the selling of clothing, home products and luxury food products. In 2012, M&S announced a series of measures designed to reduce the use of harmful substances in its textiles supply chain. It released a new version of Environmental and Chemical Policy, which imposes standards on its textiles suppliers and also features a series of commitments designed to accelerate the development of less harmful chemicals use in textile production and dyeing process. As a result, processes were being used in the M&S supply chain that reduced the environmental impact such as Cold Batch Dyeing, a process that uses 50 % less water and reduces carbon by 30 % [39].

Green purchasing can be considered as one of the major processes of GSCM [40]. It takes account of organizational sourcing decision with a focus on reducing use of environmentally unsustainable materials by developing purchasing policy, defining environmental objectives, and monitoring performance of suppliers [41, 42]. Production of environmentally friendly products requires substantial efforts to be made toward green procurement. For example, Tyson Foods requests its pork suppliers to house pigs in environments that "allow sows of all sizes to stand, turn around, lie down and stretch their legs." Such request was made due to pressure from groups such as the Humane Society and Green Century Capital Management, which called upon the food industry to terminate the use of highly constraining, two-feet-wide gestation crates. Furthermore, Tyson indicated to its suppliers that it would be making greater use of the third-party auditing services that leverages to ensure quality and practice standards with minimal impact to the environment [43].

1.2.3 Extended Producer Responsibility

Different from the notion of corporate environmental management and green supply chain management, extended producer responsibility (EPR) is referred as management practices including take-back, recycling, and final disposal of products that are helpful for manufacturing enterprises to relieve the environmental burdens of their products [40]. EPR focuses on utilizing reusable materials and components by incorporating modular design and capturing residual values from returned products. In comparison to the environmental management standard on ISO 14000 which is about process control with environmental consideration, EPR is concerned with the management practices by manufacturing enterprises on product take-back, recycling, and final disposal to reduce harms caused by their products to the environment [40, 44].

For example, General Mills, an American food company, is being asked by its shareholders to fund recycling programs for its packaging waste. They demand General Mills' brands such as Cheerios, Pillsbury, Yoplait, and Green Giant to report on their efforts in managing post-consumer product packaging. As a result, General Mills establishes extended producer responsibility programs to take responsibility for the disposal of its products and packaging by running take-back programs and funding collection efforts. Using recycled material in place of virgin material typically results in fewer carbon dioxide emissions. One salient benefit of extended producer responsibility is that it drives General Mills to invest in redesigning products and packaging to be smaller, lighter, and easily recyclable [45].

1.2.4 Summary

Environmental management helps firms sustain economic viability without depleting environmental resources. Environmental sustainability can be achieved with efforts across supply chain members, beginning with intra-organization corporate environmental management, and continuing across firms efforts all phases of a supply chain. Supply chain sustainability is a business issue affecting an organization's supply chain and logistics network in terms of environmental, risk, and waste costs. There is an increasing quest for integrating environmentally sound techniques and practices into supply chain management. A sustainable supply chain facilitates partner firms to seize the opportunities of value creation and offers prominent competitive advantages. With potential financial gains, EPR encourages firms to design environmentally friendly products and to be responsible for the costs of managing their products at end of life.

1.3 Environmental Disclosure

1.3.1 Environmental Disclosure

Firms inform the public of their financial performance, for better or for worse. The rise of corporate social responsibility has established the foundation for transparency and accountability in environmental and sustainability performance of organizations [46]. The rise of importance of these disclosures consequently led more firms to communicate with their publics of their environmental performance information. This information plays a role in shaping public perception and environmental reputation.

1.3.2 Definition of Environmental Disclosure

Environmental disclosure is information items produced by an organization—whether it is in the past, present or future—in relation to its environmental management activities and performance [47].

1.3.3 Types of Environmental Disclosure

A study developed a content analysis index based on the Global Reporting Initiative (GRI) sustainability reporting guidelines to assess the extent of discretionary disclosures in environmental and social responsibility reports [48]. Launched in 1997, the GRI s a leading organization in the sustainability field. GRI promotes the use of sustainability reporting as a way for organizations to become more sustainable and contribute to sustainable development [49].

The content analysis index—a scoring model containing 95 line items—was developed based on the GRI to measure the level of discretionary environmental disclosures in environmental and social responsibility reports or similar disclosures provided on the firm's website. This index consisted of 7 categories, dividing 4 items to represent "hard" disclosure and 3 items as "soft" disclosure.

1.3.3.1 Hard Disclosure

The study defines hard disclosure as economic based and paper based annual report. There are 4 main items that falls into this category:

– Disclosures relating to a firm's environmental protection and management systems (e.g. informing stakeholders of ISO 14001 implementation)
– Credibility of a firm's environmental report disclosures (e.g. firms with independent verification of environmental reports, products and environmental programs certified by independent agencies and external parties)
– Disclosure of specific environmental performance indicators (e.g. actual pollution emissions, conservation and recycling efforts as well as historical environmental protection trends of the firm such as targets and industry average)
– Disclosures of environmental capital spending

Hard environmental disclosure is information that is factual and objective. This makes it relatively difficult for poor environmental performers to imitate.

1.3.3.2 Soft Disclosure

The study also defines that soft disclosure as information on website and press releases which are more incentive consistent than economic based disclosure in the following 3 categories:

– Disclosure of vision and environmental strategy claims (e.g. informing the public of the management's effort and commitment towards protecting the environment)
– A firm's environmental profile under the existing and upcoming environmental regulations
– Disclosures of environmental initiatives (e.g. environmental management training for employee, environmental accidents contingency plans, internal environmental appraisals and audit, and community involvement through sponsorship)

Soft environmental disclosure also reflects true environmental commitments of a firm however these practices can easily be mimicked by other companies.

1.3.4 Case Examples of Environmental Disclosure

1.3.4.1 Royal Dutch Shell

Royal Dutch Shell takes a hard disclosure approach. Royal Dutch Shell has been submitting the firm's environmental reports to external auditors for verification since 1997. The firm strives to develop verification methods and internal systems for reliable data collection and became a leader in the world of green reporting [50]. The firm now reports on namely 5 different categories of environmental protection, including sustainable development, environment, society, safety and performance data [51].

1.3.4.2 Bristol-Myers Squibb

In comparison, Bristol-Myers Squibb, a health and personal care products organization, takes a soft disclosure approach. The organisation uploads and updates frequently the structured environmental report on its corporate website. The report includes current sustainability issues, sustainability goals and performance indicators, stakeholder engagement et cetera—a wide range of topics are covered, from energy use to philanthropic sponsorships of fighting AIDS/HIV in Africa [52].

1.3.5 Public Perception of Environmental Disclosure

"Greenwashing" is a tactic where environmental disclosure is used to obscure the public's perception of how "green" the firm is rather than actual efforts in environmental protection [53]. Greenwashing, which concerns to the validity and falsity of environmental advertisements, can be categorized in 4 categories:

1. Ambiguous: containing statements or phrases with claims that are too broad to have a clear meaning (e.g. Phrases such as "ozone-friendly," "Good for the environment," or "All-natural," in describing the benefits of their products.)
2. Omission: Leaves out information that is important in evaluation of its truthfulness or reasonableness (e.g. Claimed in its advertising that a particular trash bag was biodegradable, but failed to tell consumers that these bags would not degrade unless with sufficient sunlight and water)
3. Fales: Claims that are clearly lies or misleading (e.g. A large oil company advertising that its unleaded gas causes no pollution)
4. Acceptable: Making truthful environmental claims in a justifiable and clear way [54].

"Greenwashing" is so greatly opposed by the public that the University of Oregon in partnership with EnviroMedia Social Marketing has created the Greenwashing Index. This website allows the public to upload and rate how "bogus" the advertisement is considered to be [55]. The Greenwashing Index strives to educate consumers on how to evaluate a green advertisement and encourage them to determine if they are witnessing greenwashing. Consequently with a greenwashing savvy public, organizations would be mandated to "have a sustainable business before they advertise [that] they're a sustainable business" and "be accountable for sustainable practices they claim to have" [56].

From the general public's point of view, there is a thin line between genuine philanthropic green advertising and "greenwashing". In January 2007, Diesel—an

Italian fashion design company—launched the "Global Warming Ready" advertising campaign. Partnering with StopGlobalWarming.org, an online grassroots initiative designed to bring together voices to demand solutions to Global Warming, Diesel intended to "grab and hold people's attention a little longer than a news feature can, make them think twice about the consequence of all our actions and realize our individual responsibility." [57].

The "Global Warming Ready" campaign depicted surreal scenes of the post-Global Warming world. London became an island, New York was underwater, Paris turned into a jungle, and the once Nordic Finland turned into a desert and so forth. These striking images evoked mixed reviews. On the forum of Ads of the World—an advertising archive and community owned by WebMediaBrands—there were divergent comments on the campaign. Credit was given to the picturesque imagery, however some audience was doubtful about the motives behind this campaign since global warming was such a hot topic at that time. Some consumers were having difficulties trying to link the two messages—the Diesel new season collection and global warming—together [58]. However, the actual collection was not intended to have any connection to global warming. Diesel aimed to use the surreal and striking images of this campaign was to draw people's attention and raise awareness on global warming. And it did—whether the public liked it or not—the tongue-in-cheek campaign did draw attention towards the issue.

A research study modeled the impact of communicates in three distinct stages: lead generation, appointment conversion and sales closure [59]. It establishes that a series of Integrated Marketing Communications (IMC) can bridge the gap between a disconnection between the marketing and sales functions of many organizations.

Empirical evidence of interactive and carryover effects across media sources in the lead generation stage. Showed that communications spending can directly contribute to delays in sales force follow-up, linking media spending more directly to the sales process. This shows that through marketing communications, customers can better understand organizational products and practices and sales leads are triggered.

Environmentally conscious customers would take into consideration if a firm holds similar environmental protection beliefs as well as the intrinsic "feel good factor" for them when using an environmentally friendly product. Green advertising emphasizes rigorous marketing efforts across various marketing functions, such as promotion and product development, communicating messages (e.g. product characteristics and their environmental impact) with the aim of reducing the uncertainty of customers when deciding to make a particular purchase.

1.4 Environmental Reputation

1.4.1 What Is Green Advertising?

Every day we are exposed to and somewhat bombarded by advertisements telling us about the new must have products—ranging from electric cars to vintage (recycled) clothing. More recently, however, with the rising awareness of environmental sustainability of the general public, many organizations seek to "green" their public images.

Green advertising is a marketing tool to inform and publicize information related to the environmental efforts and commitment of an enterprise [60]. Disclosing this information can take place in many forms, including—but not limited to—a corporation's annual report, sponsoring environmental interest groups, promoting environmental information about products, providing environmental information on corporate websites to publicize the environmental actions of the firm and so forth. Green advertising is considered marketing activities of organizations intended to promote the environmental qualities of business operations and processes [61]. By letting the general public know about their environmental efforts, commitment and development, the needs of the environmentally conscious customer are in turn gratified.

Organization discloses information relating to environmental issues involving details such as energy usage, environmental expenditure, and carbon emission. For example, Mitsubishi Electric, a multinational electronics and electrical equipments manufacturing company headquartered in Tokyo, Japan, continuously inform the public on their environmental objectives and performances through reports on their corporate website. They also contribute to environmental education by running an informative website aimed at elementary school students [62], the website uses fun and interactive ways to teach children about environmental protection. Additionally, Mitsubishi Electric runs a series of outdoor classrooms held by different offices in the country. Staff from each office would plan and organize a field trip for children in their surrounding area. For example, on October 6, 2012, they collaborated with the Society for Nurturing Cherry Blossom Trees in Zugaike Park near the firm's High Frequency & Optical Device Works. This collaborative event involved pruning of cherry trees and cleaning the park and taught both adult and children participants the difficult nature of conservation work [63].

Currently, Mitsubishi Electric is working towards a low-carbon, recycling-based society. One of the efforts includes seeking a new energy potential with a wireless sensor powered by small vibrations. In 2012, the firm developed a sensing system that generates electric power with the minor vibrations that occur generally to wirelessly transmit data collected using a sensor [64]. This innovation unleashes new potential for the search of environmentally friendly energy sources.

Through marketing communications such as press releases, sponsorship of environmental interest groups and publication of corporate documents, enterprises can take advantage of green advertising to inform, incite and remind customers about their environmental stance [65]. An example is Danone, a French food-products multinational corporation. One of their sustainability goals is to bring healthy food to as many people as possible. Effort towards this goal has prompted global sales growth of 50 % from 2007 to 2012, or 600 million to 900 million people. This was all because of accessible pricing and distribution in emerging countries. Danone continues to support the Livelihoods Fund to benefit poor rural communities, and also educating communities on health and wellness [66].

Green advertising can be categorized into 3 groups [67]:

Advertisement	Examples
Directly or indirectly addressing the relationship between a product/service and the natural environment	Procter & Gamble "Tide Coldwater" campaign – Launches Tide Coldwater specially designed to clean in lower temperatures saving energy – One of the PR initiatives, the ColdWater Challenge asks consumers to wash their laundry in cold water to reduce energy use – Tide Coldwater also donated $100,000 to the National Fuel Funds Network The far-reaching campaign included national advertising, in-store programs, product sampling, a strong internet presence, consumer promotions and strategic alliances and promoted energy saving at the same time [68]
Promote an environmentally responsible lifestyle with or without highlighting a product/service	HSBC "No Small Change" U.S. retail banking campaign – Offered advice for consumers to reducing their carbon footprint – Helped to find green consumer products to give as an incentive to customers who signed up for an paperless banking offering – Helped to secure events during Earth Day that the bank could sponsor in New York City The campaign reached 103 % of its new account goal, and won one of the first three Green Effies for Advertising Effectiveness [69]
Present an image of corporate environmental responsibility	GE "Ecomagination" campaign – Long running and multimedia campaign to promote GE's green position and efforts in environmental position. – Including information on projects such as fuel efficient freight locomotive as well as captivating images of wind technology on Instagram This campaign strives to display the organizations' "commitment to build innovative solutions for today's environmental challenges while driving economic growth." [70]

1.4.2 Role of Green Advertising

There are various green advertising approaches. Issuing environmental reports encourages companies to be more meticulous about environmental performance reducing environmental risk, and creates good public relations [50]. Sponsoring environmental interest groups, representing corporate environmental responsiveness, also develops public relations [71]. Launching marketing campaigns with a focus on the eco-friendliness of products can be useful for manufacturers to gain legitimacy and environmental reputation.

For example by communicating through green product attributes of the electric, zero-emissions vehicle—the Nissan Leaf—a green halo effect was brought onto the reputation of Nissan. Gaining the company 5th place on the Interbrand 2013 Best Green Brands ranking, jumping from 16 positions in comparison to the previous year. Nissan also obtained a relatively low gap score of +2.06, which means the actual environmental performance and environmental perception of customers were akin. They have also achieved remarkable work on increasing disclosure of policies, programs, and impacts [72].

The Leaf went from an environmental angle from the beginning. Roel de Vries —corporate vice president and global head of marketing, communication and brand strategy of Nissan—told GreenBiz that the first advertisement featuring a polar bear "was a very clear link between the environment and the impact of cars on the environment, and the fact that we are bringing a car to market that doesn't have any emissions. So, it was a statement of the fact that cars don't have to be bad for the environment. At that stage, we didn't talk so much about the specifications of the car or its pricing. We very much went with the concept of zero emissions." [73].

The communication theory establishes that customer will interpret advertising information with respect to its meanings, make judgements and attributions about the intention and consequently take that into account and decide on their actions towards the firms [74]. A research study has found that environmentally friendly conscious consumers may refrain from purchasing from a particular brand if they believe it does not agree with their beliefs [75]. Hence green advertising is advantageous in influencing and moulding stakeholders' perceptions [76]. Through showcasing their environmentally friendly achievements, differentiating their environmental direction from the competition [77] as well as establishing a positive green image.

However depending on different situational factors, green advertising can bring adverse effects to a manufacturer. As aforementioned, "greenwashing" is an issue that the public is aware of. Vigorous publicity on environmental performance can be viewed as a questionable attempt to manipulate customers' impressions. Moreover, if advertising efforts on environmental protection were found greater than actual efforts in environmental protection practices then it would be a scandalous hypocrisy at a very high public relations price.

The balance of promotional efforts needs to be handled carefully. Manufacturers with good environmental protection are expected to reduce their environmental damage, and it is an almost imbedded in consumers' expectations. However the process of revealing these practices—promotional efforts to highlight the environmental benefits of products and manufacturing processes—might be found excessive. This would raise customers' queries in questioning the companies' intentions. Skepticism would be raised about the environmental claims [78] and might be not contribute towards favorable financial results.

1.4.3 Divergent Views on Green Advertising

Ever since its debut in the 1980s, green advertising has drawn immense attention from brands, advertisers and researchers and the same time. Green advertising can create both affective and cognitive benefits among consumers [79]. An example of affective benefit is increasing consumers' satisfaction by purchasing an environmentally friendly product; a research study has indicated that environmental labels can positively influence customer perceptions of a brand's environmental performance [75]. In terms of cognitive benefits, green advertising can contribute to establishing social norms. A research study conducted in 2008 investigating consumer recycling behavior has found that consumers are likely to be affected by peers. If a consumer knows that the majority of other consumers recycle bottles and avoid buying non-recyclable ones, that consumer is more likely to do the same [80].

While a firm's green advertising efforts can be attributed to altruistic motives [81], it can also be perceived as self-presentation and exercises in public relations. The general public is weary and concerned about "greenwashing". It is only human nature to be skeptical, especially when we are facing uncountable new [advertising] claims everyday whether one likes it or not. Consumers—green consumers in particular—are skeptical of promotional messages emphasizing the environmental image of enterprises. Green consumers are receptive to green advertising, however they also tend to be careful shoppers who vigilantly evaluate information on products, including that from advertising. Consequently, when implementing green advertising, marketers should be cautious of using ambiguous or misleading messages [82].

For example in 2007, the Advertising Standards Authority (ASA)—the independent regulator of advertising in U.K.—ruled against Shell on a press advertisement. The advert depicted an image of refinery chimneys emitting out flowers with the slogan, "Don't throw anything anyway. There is no away." Environmental group, Friends of the Earth complained to the ASA that the image misrepresented the environmental impact of the organization's core activities. The ASA ruled the wording of advertisement implied that a significant amount of Shell's emissions were recycled to grow flowers, where in reality, it was only a very small amount in comparison to the global actions of the organization [83].

Taiwanese electronics manufacturers are key players in supplying a wide spectrum of electronics parts and components worldwide. They are also pioneers in research and development for information technology production [84].

Additionally, Taiwan also plays a large role in industrial environmental protection. Taiwan implements various industrial environmental improvement programs. For example the Industrial Waste Minimization and Green Design Network programs have been running since the 1980s with a key focus to encourage environmentally conscious manufacturing. This industry has strived to make an effort at environmental protection through greening of their manufacturing processes, with reduced consumption of resources and improved product design to facilitate end-of-life product management [85].

A study based on Taiwanese electronics manufacturers explored the value of green advertising in sharing and publicizing information about organizational achievements in environmental protection in a business-to-business (B2B) context [86]. The study examined 122 electronics manufacturers listed in the Taiwan Stock Exchange Corporation (TWSE) market and the Gre Tai Securities Market (GTSM) in Taiwan. These electronics manufacturers produce a variety of technological products, ranging from semiconductors, optoelectronics, computer and peripheral equipments, to electronics components.

From the study based on Taiwanese electronics manufacturers, the findings demonstrated that there is a substantive value of green advertising for manufacturers to acquire a favorable environmental reputation. Manufacturers should use green advertising to publicize their environmental achievements to strengthen the relationship between environmental management practices and environmental reputation.

1.4.4 Green Advertising Limitations

Investors may not be entirely convinced by the effectiveness of green advertising. Investors have reservations about corporate green marketing activities. A research study conducted in 2000 has found that, measuring a firms financial performance by growth in earnings per share, firm size and the advertising-to-sales ratio, investors only seem to be more comfortable with green marketing activities by those with better financial performance. Whereas for financially weaker firms, investors consider green marketing activities as opportunistic [87].

Strong voices in society, such as environmental activists and environmental interest groups have apprehension about corporate deception. Under the scrutiny of environmentalist groups, firms may elicit unwanted attack and attention from these groups, questioning the effectiveness of organizational initiatives or the intention behind their environmental actions [88]. If this issue were not handled properly, detrimental effects would be brought upon the firm's reputation.

For example, the United Kingdom Advertising Standards Authority found that a series of television advertisements by the Malaysian Palm Oil Council was misleading. Claiming that the industry was good for the environment, one of the advertisements—which appeared on television in Europe, Asia and the U.S.—featured a man jogging through a rain forest and shots of palm-oil plantations with a voice-over saying "Malaysia palm oil. Its trees give life and help our planet breathe," [89]. However environmentalists revealed the truth that in Indonesia, where Malaysian palm-oil companies hold large operations, plantation development was destroying the natural habitat. The U.K. authority ruled that the advertisements were misleading the public. In addition, environmentalists escalated attacks on palm oil, declaring it a major driver of forest loss. Consequently the Malaysian Palm Oil council had to address to the TV advertisements. Yusof Basiron—chief executive of the palm oil council—told Wall Street Journal: "We decided it was about time we gave a public-service announcement to the consumer," and salvage the situation and re-promote the green credentials of the industry by hiring TBWA worldwide.

1.4.5 Definition of Environmental Reputation

Corporate reputation involves collective judgments from the customers' perspective, judging on a firm's corporate performance relative to those of competing companies over time [90]. It is also vital for manufacturers to establish and develop an environmental reputation with the public, revealing their environmental commitment as well as future prospects to sustain these commitment efforts In developing this reputation, a manufacturer should show their achievements in environmental protection which convince the customers that their purchases are produced in an environmentally friendly way [90]. Apart from the environmental management practices, responsiveness in sharing news on environmental initiatives may also affect the customers' impression about a manufacturer.

According to the resource-based view, reputation is an intangible asset that explains competitive advantage in terms of inimitable qualities. Environmental reputation is a quantitative and qualitative measure and collective judgments from the consumers on a firm's environmental activities as disclosed in annual reports, advertising, company websites et cetera. Intuition and theoretical relations between the intangible resource of the firm and the ability of the firm to offer genuine qualitative signals to the most powerful stakeholder groups, also suggest the qualitative nature of environmental disclosure is more likely to enhance the environmental reputation of the firm [91].

1.4.6 Environmental Legitimacy and Environmental Reputation

Environmental legitimacy can be defined as the generalized perception or assumption that a firm's corporate environmental performance is desirable, proper, or appropriate [92]. Three primary forms of legitimacy have been identified: pragmatic, based on audience self-interest; moral, based on normative approval: and cognitive, based on comprehensibility and taken-for-grantedness. Stakeholders of a firm evaluate the firm's legitimacy according to their own beliefs, norms and logical decisions.

Through analyzing media reports and stock prices of 100 firms over a five-year period, a study shows that environmental legitimacy is related to unsystematic stock market risks, this gives managers incentives to manage environmental reputation with positive information released to the media [93]. To determine environmental legitimacy of a firm, investors would seek information about the firm from sources such as the media and other informed parties including other stakeholders, government regulators, suppliers, customers and competitors. Managers of firms with low environmental reputation should manage the firm's environmental performance to minimize negative media reports and the need to disclose environmental liabilities, through green advertising.

In the case when environmental management practices are valuable for attracting customers to purchase environmentally friendly manufactured goods, manufacturers need to exhibit their achievements in environmental protection in order to develop environmental reputation. For example, a study based on Taiwanese electronics manufacturers explores the value of green advertising in sharing and publicizing information about organizational achievements in environmental protection in a business-to-business (B2B) context [86].

This study also sheds light on green advertising in the business-to-customer (B2C) context. The study found that when green advertising is used to publicize information about the environmental efforts of manufactures, environmental management practices are positively related to environmental reputation. Green advertising is required in order to acquire an environmental reputation. Environmental management practices based on environmentally conscious manufacturing and product development alone do not contribute to the environmental reputation of firms.

According to the communication theory, the study demonstrated the important role of green advertising for environmental reputation. Disseminating information related to corporate environmental protection efforts better inform customers about the environmental impact of organizational products and processes, and thus generating a positive environmental reputation.

1.4.7 Environmental Indices

Environmental reputation can be a valuable asset for an organization to gain legitimacy and customer support through communication of their environmental achievements. A creditable environmental reputation is a valuable intangible asset that is largely based on customers' perceptions of the firm. Managers should also be aware that openness in sharing relevant environmental information as well as implementation of environmental management practices could influence customers' judgments and their opinions about their firms.

Three central areas of concerns in measuring the sustainability and ethical impact of companies or businesses are: Environmental, social and corporate governance (ESG). To establish environmental legitimacy, firms can follow different environmental indices to be on par with industry norms.

1.4.7.1 MSCI ESG Indices

The MSCI ESG Indices is an index provider with it's own ESG research team. They strive to allow clients to more effectively benchmark ESG investment performance, issue index-based ESG investment products, as well as to manage, measure and report on their compliance with ESG mandates. Under MSCI ESG research there is the MSCI Emerging Markets (EM) ESG Index, a capitalization weighted index that allows companies with high ESG performance to see their relative scores to their sector peers. Under the MSCI SRI Indices there are the MSCI Global SRI Indices and MSCI KLD 400 Social Index, which target companies with high ESG ratings relative to their sector peers and includes 400 (large, mid and small capitalization) companies with high ESG ratings relative to the constituents in the MSCI USA Investable Market Index respectively [94].

1.4.7.2 Dow Jones Sustainability Indices

Launched in 1999, the Down Jones Sustainability Indices are offered cooperatively by RobecoSAM and S&P Dow Jones Indicies. Tracking the stock performance of global leading companies on economic, environmental and social criteria, the indices serve as benchmarks for sustainability considerations. Providing investors and managers a guide for sustainable practices [95].

1.4.8 Role of Environmental Reputation

Environmental reputation takes on a mediating role on the relationship between environmental management practices and financial performance. To the public, manufacturers practicing environmental management show that they are environmentally conscious and strive to mitigate their environmental harms. To customers who care about manufacturing-related pollution, this exemplifies a positive environmentally friendly image [93]. Hence, a positive reputation for environmental protection is built.

There are increasing findings that indicate enterprises with an established environmental reputation achieve better financial performance. Environmental reputation is a valuable asset to draw in investors, in negotiation with financers as well as building customer satisfaction and loyalty [96]. For example, the 2013 CSR RepTrak® 100 study results show that 73 % of the 55,000 consumers surveyed are willing to recommend companies perceived to be delivering on Corporate Social Responsibility. The findings showed that The Walt Disney Company has the best perception for citizenship. Approximately 50 % of consumers surveyed agree that Disney is a good corporate citizen that supports good causes and protects the environment [97]. This study shows that positive reputation can gain financial benefits in terms of sales. Additionally, a study based on Taiwanese electronics manufacturers also found that when manufactures have a favorable environmental reputation, they achieve better financial performance. It is beneficial for manufacturers to develop an environmental reputation to obtain bottom-line gains [98].

However, for those manufacturers already with an established environmental reputation, green advertising does not contribute to improve bottom-line gains. This can be explained by two causes. First, the market may become skeptical about corporate motives for engaging in green advertising effort, subsequently perceives a negative image of manufactures [99]. Second, the general environmental protection perception of electronic manufacturers is often unfavorable, with constant accusations for being a highly polluting industrial sector.

In the case where manufacturers have an established environmental reputation, customers expect them to be environmentally responsible regardless. Subsequently would question the intention behind their green promotional efforts. In this scenario green advertising exerts negative effects on the environmental reputation and financial performance relationship. Green advertising can be damaging to manufacturers that already have a good environmental protection reputation.

The results highlight the importance of reputation in strengthening the link between environmental management practices and financial performance. Based on the communication theory, the role of green advertising was identified as an important element to leverage the environmental manufacturing and product stewardship efforts to achieve an environmental reputation. Manufacturers should publicize information conveying their environmental commitment and efforts, which may not be visible to the market.

A good environmental reputation fosters a good foundation for manufacturers to reinforce environmental responsibility in accordance with customers' expectations, support their continuous efforts to reduce environmental harms, and encourage customers' continuous support and patronage. So firms with better environmental reputation are more likely to achieve better financial performance with environmental management practices.

1.4.9 Executive Summary

The beneficial and detrimental effects of green advertising have been outlined. In publicizing information related organizational efforts on environmental protection, relationship between environmental reputation and financial gains and the importance of environmental management practices, there are the following aspects to consider:

1.4.9.1 Green Advertising

There are different circumstances which managers should adopt green advertising, particularly when reputational and financial rewards are objectives of the firm. Green advertising makes great impacts when manufacturers implement environmental management practices. Managers should consider using green advertising to publicize the firm's environmental commitment and efforts. This approach allows the firm to acquire environmental reputation and subsequently improve financial performance. However the effect of green advertising on financial performance can be counterproductive. Managers need to bear in mind that green advertising is detrimental to those firms who are already renowned for their environmental protection.

1.4.9.2 Environmental Reputation and Financial Performances

A positive environmental reputation is a valuable intangible asset of a firm that reflects successful fulfillment of stakeholders' expectations [100]. For instance, Foxconn Technology group, the largest electronics manufacturing service provider listed among the Fortune Global 500 companies and the largest corporation in Taiwan, has been working closely with the government. With their efforts to promote energy conservation, Foxconn was recognized with the 'Model of an Energy Saving Corporation' award. Stakeholders deem such positive environmental reputation favourable, subsequently attracting customers and investors, leading to financial rewards.

1.4.9.3 Importance of Environmental Management Practices

Managers should consider environmental management practices as an opportunity to improve performance while protecting the environment at the same time. During manufacturing processes and product development, managers can implement practices such as minimizing pollution and consumption of resources, substituting harmful materials, reducing waste, adopting new technologies and controlling carbon emissions. This would improve environmental reputation and bring about improved financial performance.

References

1. Antweiler W (2014) Elements of environment management. University of Toronto Press, Toronto
2. Sarkis J (2014) Green supply chain management. Momentum Press, New York
3. Lai K-H, Wong CWY (2012) Green logistics management and performance: some empirical evidence from Chinese manufacturing exporters. Omega 40:267–282

4. van Hemel C, Cramer J (2002) Barriers and stimulus for eco-design in SMEs. J Clean Prod 11:439–453
5. Zhu QH, Sarkis J, Cordeiro JJ, Lai KH (2008) Firm-level correlates of emergent green supply chain management practices in the Chinese context (International Journal of Management Science). Omega 36(4):577–591
6. Guide VDR, Teunter RH, Van Wassenhove LN (2003) Matching demand and supply to maximize profits from remanufacturing. Manufact Serv Oper Manag 5:306–318
7. Lai K-H, Wong CWY, Cheng TCE (2012) Ecological modernisation of Chinese export manufacturing via green logistics management and its regional implications. Technol Forecast Soc Chang 79:766–770
8. Klassen RD, Whybark DC (1999) Environmental management in operations: the selection of environmental technologies. Decis Sci 30:601–631
9. Shrivastava P (1995) Environmental technologies and competitive advantages. Strateg Manag J 16:183–200
10. Hilmola O (2011) North European companies and major Eurasian countries—future outlook on logistics flows and their sustainability. Int J Shipping Transport Logistics 3:100–121
11. Lai K-H, Lun VY, Wong CWY, Cheng TCE (2011) Green practices in the shipping industry: conceptualization, adoption, and implications. Resour Conserv Recycl 55:631–638
12. Li S, Shi L, Feng X, Li K (2012) Reverse channel design: the impacts of differential pricing and extended producer responsibility. Int J Shipping Transport Logistics 4:357–375
13. Clancy H (2014a) Why Philips' EcoDesign play is paying off in more than one way. Available from http://www.greenbiz.com/blog/2014/07/07/why-philips-wants-product-designers-think-circular
14. Wong CWY, Lai K-H, Shang K-C, Lu C-S, Leung TKP (2012) Green operations and the moderating role of environmental management capability of suppliers on manufacturing firm performance. Int J Prod Econ 140:283–294
15. Rabin E (2003) ISO 14000: a worldwide blueprint for environmental management? Available from http://www.greenbiz.com/news/2003/04/27/iso-14000-worldwide-blueprint-environmental-management
16. Makower J (2012) Color It green: Nike to adopt waterless textile dyeing. Available from http://www.greenbiz.com/blog/2012/02/07/color-it-green-nike-adopt-waterless-textile-dyeing
17. Sbihi A, Eglese RW (2007) Combinatorial optimization and green logistics. 4OR 5:99–106
18. Staff G (2009) Logistics efficiencies show JohnsonDiversey that less is more. Available from http://www.greenbiz.com/news/2009/07/10/logistics-efficiencies-show-johnsondiversey-less-more
19. Sroufe R (2003) Effects of environmental management systems on environmental management practices and operations. Prod Oper Manage 12:416–431
20. Prajogo D, Tang KYA, Lai K-H (2014) The diffusion of environmental management system and its effect on environmental management practices. Int J Oper Prod Manage 34(5):565–585
21. Boiral O (2007) Corporate greening through ISO 14001: a rational myth? Organ Sci 18(1):127–162
22. Klassen RD, Vachon S (2003) Collaboration and evaluation in the supply chain: the impact on plant-level environment investment. Prod Oper Manage 12(3):336–352
23. Corbett CJ, Kirsch DA (2001) International diffusion of ISO 14000 certification. Prod Oper Manage 10(3):327–342
24. Yang J, Wong CWY, Lai K-H, Ntoko AN (2009) The antecedents of dyadic quality of performance and its effect on buyer–supplier relationship improvement. Int J Prod Econ 102(1):243–351
25. Yang J, Wang J, Wong CWY, Lai K-H (2008) Relational stability and alliance performance in supply chain. Omega 36(4):600–608

26. Koufteros XA, Nahm AY, Cheng TCE, Lai KH (2007) An empirical assessment of a nomological network of organizational design constructs: from culture to structure to pull production to performance. Int J Prod Econ 106(2):468–492
27. Koufteros XA, Cheng TCE, Lai KH (2007) "Black-box" and "gray-box" supplier integration in product development: antecedents, consequences and the moderating role of firm size. J Oper Manage 25(4):847–870
28. Clancy H (2014B) Inside panasonic's B2B energy play. Available from http://www.greenbiz. com/blog/2014/04/16/under-panasonics-eco-mantra-solutions-trump-products
29. Nicol S (2007) Policy options to reduce consumer waste to zero: comparing product stewardship and extended producer responsibility for refrigerator waste. Waste Manage Res 25(3):227–233
30. Morrow D, Rondinelli D (2002) Adopting corporate environmental management systems: motivations and results of 1SO 14001 and EMAS. Eur Manage J 20(2):159–171
31. Welford RJ (1998) Corporate environmental management, technology, and sustainable development: postmodern perspectives and the need for a critical research agenda. Bus Strat Environ 7(1):1–12
32. Hering G (2014) Boeing sees 'green diesel' in the future of air travel. Available from http:// www.greenbiz.com/blog/2014/01/15/boeing-sees-green-diesel-sky
33. Coughlin C (2013) Best buy: thinking outside the big box. Available from http://www. greenbiz.com/blog/2013/06/02/best-buy-thinking-outside-big-box
34. Sabri EH, Shaikh SN (2010) Lean and agile value chain management: a guide to the next level of improvement. J. Ross Publishing, Richmond
35. Harland CM (1996) Supply chain management, purchasing and supply management, logistics, vertical integration, materials management and supply chain dynamics. In: Slack N (ed) Blackwell encyclopedic dictionary of operations management. Blackwell, UK
36. Sarkis J, Zhu QH, Lai KH (2011) An organizational theoretic review of green supply chain management literature. Int J Prod Econ 130(1):1–15
37. Staff G (2009) IBM extends green IT offerings to supply chain management. Available from http://www.greenbiz.com/news/2009/10/20/ibm-extends-green-it-offerings-supply-chain-management
38. Guevarra L (2010) Kaiser applies new green scorecard to $1B medical. Available from http:// new.greenbiz.com/news/2010/05/04/kaiser-applies-new-green-scorecard-medical-supply-chain
39. Murray J (2012) M&S cleans up supply chain with hazardous chemical commitments. Available from http://www.greenbiz.com/news/2012/10/25/ms-cleans-supply-chain-hazardous-chemical-commitments
40. Lai K-H, Wong CWY, Lun YHV (2014) The role of customer integration in extended producer responsibility: a study of Chinese export manufacturers. Int J Prod Econ 147:284–293
41. Chen CC (2005) Incorporating green purchasing into the frame of ISO 14000. J Clean Prod 13:927–933
42. Wu SJ, Melnyk SA, Calantone RJ (2008) Assessing the core resources in the environmental management system from the resource perspective and the contingency perspective. IEEE Trans Eng Manage 55(2):304–315
43. Strategic Sourceror (2014) Food companies put green procurement
44. Subramanian R, Gupta S, Talbot B (2009) Product design and supply chain coordination under extended producer responsibility. Prod Oper Manage 18(3):259–277
45. Bardelline J (2011) Shareholders Urge P&G, General Mills to Step Up Package Recycling. Available from http://www.greenbiz.com/news/2011/04/28/shareholders-urge-pg-general-mills-step-up-package-recycling
46. Levy DL, Kaplan R (2008) Corporate social responsibility and theories of global governance: strategic contestation in global issue arenas. In: Andrew C et al. (ed) The Oxford handbook of corporate social responsibility. Oxford University Press, Oxford

47. Berthelot S, Cormier D, Magnan M (2003) Environmental disclosure research: review and synthesis. J Acc Lit 22:1–44
48. Clarkson PM et al (2008) Revisiting the relation between environmental performance and environmental disclosure: an empirical analysis. Acc Organ Soc 33(4–5):303–327
49. Global Reporting Initiative (2013) What is GRI?. Cited 3 Dec 2013. Available from https://www.globalreporting.org/information/about-gri/what-is-GRI/Pages/default.aspx
50. Kolk A (2000) Green reporting. Harvard Bus Rev 78(1):15–16
51. Global S (2012) 2012 Shell sustainability report. Cited 3 Dec 2013. Available from http://www.shell.com/global/environment-society.html
52. Bristol-Myers Squibb Company (2013) Sustainability. Cited 3 Dec 2013. Available from http://www.bms.com/sustainability/Pages/home.aspx
53. Beets SD, Souther CC (1999) Corporate environmental reports: the need for standards and an environmental assurance service. Acc Horiz 13(2):129–145
54. Carlson L et al (1996) Does environmental advertising reflect integrated marketing communications?: an empirical investigation. J Bus Res 37(2):225–232
55. Marketing ES (2013) Greenwashing index. Cited 3 Dec 2013. Available from http://www.greenwashingindex.com/ads/
56. Marketing ES (2013) About Greenwashing. Cited 27 Nov 2013. Available from http://www.greenwashingindex.com/about-greenwashing/
57. Warming SG (2013) Diesel. Cited 27 Nov 2013. Available from http://www.stopglobalwarming.org/partners/diesel/
58. Mediabistro Network Inc (2013) Diesel: global warming, rushmore mountain. Cited 27 Nov 2013. Available from http://adsoftheworld.com/media/print/diesel_global_warming_rushmore_mountain
59. Smith TM, Gopalakrishna S, Chatterjee R (2006) A three-stage model of integrated marketing communications at the marketing-sales interface. J Mark Res 43(4):564–579
60. Keller KL (2001) Mastering the marketing communications mix: micro and macro perspectives on integrated marketing communication programs. J Mark Manage 17(7–8):819–847
61. Kilbourne WE (1995) Green advertising: salvation or oxymoron? J Advertising 24(2):7–19
62. Mitsubishi Electric Corporation (2013) Environment—disclosure and dissemination of environmental information. Cited 27 Nov 2013. Available from http://www.mitsubishielectric.com/company/environment/report/communication/information/index.html
63. Mitsubishi Electric Corporation (2013) Environment—Mitsubishi Electric outdoor classroom. Cited 3 Dec 2013. Available from http://www.mitsubishielectric.com/company/environment/report/communication/classroom/index.html
64. Mitsubishi Electric Corporation (2013) Environment—new energy potential—a wireless sensor powered by small vibrations. Cited 3 Dec 2013. Available from http://www.mitsubishielectric.com/company/environment/ecotopics/vibration/index.html
65. Narayanan S, Manchanda P (2009) Heterogeneous learning and the targeting of marketing communication for new products. Mark Sci 28(3):424–441
66. Interbrand (2013) Danone. Cited 3 Dec 2013. Available from http://www.interbrand.com/en/best-global-brands/Best-Global-Green-Brands/2013/Danone
67. Banerjee S, Gulas CS, Iyer E (1995) Shades of green: a multidimensional analysis of environmental advertising. J Advertising 24(2):21–31
68. Gamble P (2012) P&G "Take a Load Off" campaign, together with actress Vanessa Lachey, empowers consumers nationwide to switch to cold water laundry washing this earth day. Cited 3 Dec 2013. Available from http://news.pg.com/press-release/pg-corporate-announcements/pg-take-load-campaign-together-actress-vanessa-lachey-empow
69. Consulting JO. HSBC's There's No Small Change Campaign. Cited 3 Dec 2013. Available from http://www.greenmarketing.com/clients/hsbc/
70. GE (2013) Ecomagination. Cited 3 Dec 2013. Available from http://www.ge.com/about-us/ecomagination

71. Sharma S, Pablo A, Vredenburg H (1999) Corporate environmental responsiveness strategies: the importance of issue interpretation and organizational context. J Appl Behav Sci 35(1):87–108
72. Interbrand (2013) Best global green brands 2013 nissan. Cited 3 Dec 2013. Available from http://www.interbrand.com/en/best-global-brands/Best-Global-Green-Brands/2013/Nissan
73. Makower J (2013) Leaf power: how a 'hero product' drives Nissan's reputation. Available from www.GreenBiz.com
74. Euske N, Roberts K (eds) (1987) Evolving perspectives in organization theory: communication implications. In: Jablin FM et al (eds) Handbook of organizational communication: an interdisciplinary perspective, vol 41–69. Sage Publications, Inc., Newbury Park
75. Montoro Rios FJ et al (2006) Improving attitudes toward brands with environmental associations: an experimental approach. J Consum Mark 23(1):26–33
76. Elsbach KD, Sutton RI (1992) Acquiring organizational legitimacy through illegitimate actions: a marriage of institutional and impression management theories. Acad Manag J 35 (4):699–738
77. Shrivastava P (1995) Environmental technologies and competitive advantage. Strateg Manage J 16(Special issue: Technological transformation and the new competitive landscape):183–200
78. Chan RYK (2000) The effectiveness of environmental advertising: the role of claim type and the source country green image. Int J Advertising 19(3):349–375
79. Konga Y, Zhang A (2013) Consumer response to green advertising: the influence of product involvement. Asian J Commun 23(4):428–447
80. Rokka J, Uusitalo L (2008) Preference for green packaging in consumer product choices—do consumers care? Int J Consum Stud 32(5):516–525
81. Peattie K, Peattie S (2009) Social marketing: a pathway to consumption reduction? J Bus Res 62(3):260–268
82. Shrum LJ, McCarty JA, Lowrey TM (1995) Buyer characteristics of the green consumer and their implications for advertising strategy. J Advertising 24(2):71–82
83. Tryhorn C (2007) No bouquets for Shell press ad. In: The Guardian
84. The Economist Intelligent Unit (2008) How technology sectors grow: benchmarking IT industry competitiveness 2008
85. Wong CWY, Lai KH, Shang K-S, Lu C-S, Leung TKP (2012) Green operations and the moderating role of environmental management capability of suppliers on manufacturing firm performance. Int J Prod Econ 140:283–294
86. Wong CWY, Lai KH, Shang K-S, Lu C-S (2014) Uncovering the value of green advertising for environmental management practices. Bus Strategy Environ 23:117–130
87. Mathur LK, Mathur I (2000) An analysis of the wealth effects of green marketing strategies. J Bus Res 50(3):193–200
88. Laufer WS (2003) Social accountability and corporate greenwashing. J Bus Ethics 43 (3):253–261
89. Wright T (2008) False 'Green' Ads draw global scrutiny. The Wall Street J
90. Rindova VP et al (2005) Being good or being known: an empirical examination of the dimensions, antecedents, and consequences of organizational reputation. Acad Manag J 48 (6):1033–1049
91. Toms S, Hasseldine J, Salama A (2005) Quantity versus quality: the impact of environmental disclosures on the reputations of UK plcs. Br Acc Rev 37(2):231–248
92. Suchman MC (1995) Managing legitimacy: strategic and institutional approaches. Acad Manage Rev 20(3):571–610
93. Bansal P, Clelland I (2004) Talking trash: legitimacy, impression management, and unsystematic risk in the context of the natural environment. Acad Manag J 47(1):93–103
94. MSCI Inc. (2013) MSCI ESG Indices. Cited 4 Dec 2013. Available from http://www.msci.com/products/indexes/esg/

95. RobecoSAM (2013) Dow Jones sustainability indices. Cited 5 Dec 2013. Available from http://www.sustainability-indices.com/about-us/dow-jones-sustainability-indexes.jsp

96. Milgrom P, Roberts J (1986) Relying on the information of interested parties. Rand J Econ 17(1):18–32

97. Reputation Institute (2013) 2013 CSR RepTrak® 100 Study. Cited 5 Dec 2013. Available from http://www.reputationinstitute.com/thought-leadership/csr-reptrak-100

98. Nakra P (2000) Corporate reputation management: "CRM" with a strategic twist? Publ Relat Quarterly 45(2):35–42

99. Porter ME, Kramer MR (2002) The competitive advantage of corporate philanthropy. Harvard Bus Rev 80(12):56–69

100. Freeman RE (1984) Strategic management: a stakeholder approach. Pitman Press, Boston

Chapter 2
Environmental Management Practices with Supply Chain Efforts

Abstract This chapter aims to provide a comprehensive understanding of the sustainability impacts of supply chain and studies the commonly-used environmental management practices of achieving supply chain sustainability including product and process stewardship, green services, and green logistics management. From the life cycle perspective, all of products, processes, and services, involve multiple stages which incur energy consumption, wastes, and pollutant emissions. Product stewardship is concerned with the environmental impact of products relating to their packaging design, production, delivery, usage, and development. Process stewardship is concerned with reducing adverse environmental impact in the processes ranging from production, distribution, to end-of-life product management. Green services are the services that are designed to have a lesser effect on environment and human health. Moreover, green logistics management which can be driven by economic pressure, customer pressure, and environmental regulation, helps mitigate the environmental footprints of forward and reverse flows of products, information, and services between origin and consumption. Last but not least, extended producer responsibility is a type of product stewardship that requires at a minimum the producer to be responsible for their product throughout the life cycle, which includes various approaches such as take-back programs, advanced recycling fee, and voluntary industry practices. Extended producer responsibility could improve market performance of businesses by materials recycle, waste reduction, and customer integration.

2.1 Product Stewardship

With an increasing demand for organizations to be environmentally responsible, the scope of environmental management spans from product development to management of the entire product life cycle, including eco-design, clean production, recycling, and reuse with a focus on minimizing the expenses associated with manufacturing, distribution, use, and disposal of products [1].

© The Author(s) 2016
C.W.Y. Wong et al., *Environmental Management*,
SpringerBriefs in Applied Sciences and Technology,
DOI 10.1007/978-3-319-23681-0_2

2.1.1 Definition of Product Stewardship

The product-oriented environmental management practice can be referred to as product stewardship. Product stewardship is concerned with reducing environmental burden with less use of hazardous and nonrenewable materials in products development, considering the environmental impact in product design, packaging, and material used [2]. The product should be made of non-hazardous materials, is packed in minimal or reusable packaging and consumes less energy when in use. The product is also designed for easy disassembly to facilitate reuse and recycling of parts and components with the aim to reducing environmental impact upon disposal [3]. Aimed to reducing the environmental damage arising from all product-related parts and components, product stewardship promotes recycling and reuse of product components with eco-design, and using recyclable parts and packaging [4]. An exemplar of implementing product stewardship is Samsung Electronics Co., Ltd. a South Korean multinational electronics company. Samsung has established the Eco-partner Certification in 2004 to address the European Restriction of Hazardous Substances Directive (RoHS). This certification encourages active participation and cooperation of suppliers to remove hazardous substance from products at the raw materials acquisition stage. In regards to treating end-of-life products, Samsung has an extensive mobile phone recycling program, with operations in 32 countries across Asia, Europe and the Americas. In the U.S., Samsung has recycled 1.7 million pounds of e-waste through Samsung Recycling Direct. This mobile phone recycling program allows electronic products recycling in all 50 states of the U.S. [5].

Product stewardship is often mentioned alongside with extended producer responsibility. Both practices concern management of end-of-life products, however product stewardship refers to the responsibility of every party involved in the product lifecycle, not only the producer. According to the Product Stewardship Institute, extended producer responsibility is a type of product stewardship that includes at a minimum that the producer is responsible for their product and that responsibility extends to post-consumer management of that product and its packaging [6]. Extended Producer Responsibility (EPR) in some countries, such as the European Union and South Korea, is mandatory in such context as WEEE and ELV Directives, requiring producers to be responsible from production to post-consumer management of product and its packaging. This shifts the financial and management responsibility of end-of-life products upstream to producers and not on users. EPR also encourages producers to consider environmental impacts of their products and packaging during the design stage to facilitate recovery of useful residuals for new product development. In product stewardship contrarily, constitutes that the producer is the first step and has the greatest influence over adverse environmental impacts throughout a product life-cycle. However, other stakeholders such as suppliers, retailers, and consumers are assumed to be responsible.

For example, upstream suppliers would cooperate to provide raw materials that are more environmentally friendly, whereas consumers would participate in recycling of end-of-life products.

2.1.2   Components of Product Stewardship [7]

There are 10 aspects to consider when implementing product stewardship:

1. Shared responsibility:
   All stakeholders, namely suppliers, manufacturers, retailers and customers, involved in a product lifecycle are responsible when it comes to product stewardship. Specifically, manufacturers have the responsibility and need the ability to reduce the environmental footprint of their products. By reducing use of toxic substances and designing components for reuse and recyclability, manufacturers can reduce costs, improve profitability, and foster customers' values with less environmental footprints. Besides, retailers play a salient role in influencing suppliers to behave more environmentally friendly, educating consumers on choosing environmentally safe and efficient products, and enabling consumer return of products for recycling. Real change, however, cannot be achieved without the participation of consumers. It is ultimately the consumers who make the purchase decision, use and dispose the products responsibly. Responsibility to ensure products are managed safely, in regards to supply, distribution, use, dispose and recycle is shared amongst stakeholders.

2. Lifecycle thinking:
   Firms should work towards thinking of the product lifecycle as a whole to prevent or reduce environmental risks, and to improve sustainability. For example, Samsung continues to improve eco-features of its products. In 2006, Samsung started to implement an Eco-Design Process to evaluate the eco-friendliness of its new products at the development stage [8]. Under this rating system, it classifies a premium eco-product as complying with regulation standards, qualifying for environmental marks, and is innovative in reducing environmental impacts. When evaluating the environmental impact, firms need to consider raw material extraction, material processing, transportation, distribution, consumption, reuse/recycling, and disposal. Taking a holistic assessment of the environmental impact of a product life-cycle is important as a product is likely to have different environmental impacts at different stages of its cycle. Taking aluminum as an example, the extraction and processing of aluminum have an adverse environmental impact, but the use and recycling of aluminum are relatively benign.

3. Knowledge:
   Firms need to understand the potential environmental, health, and safety risks in a product manufacturing stage as well as the environmental impact in its life-cycle may cause. This allows contingency plans to be developed early to

reduce environmental risks. The responsibility for developing knowledge of a product's hazards lay on the manufacturers who are responsible for designing their products. For example, DuPont develops a comprehensive toxicology database on its products, raw materials, and degradation products, which lays a foundation for knowledge on product safety for workers, consumers, and the ambient environment [8].

4. Supply chain communication:
   It is important to share information with upstream and downstream partners in a supply chain to allow partners to fully understand potential hazards. For example, Samsung works closely with suppliers for carrying out a series of site audits on its major suppliers in China to assess their practices in human rights, working conditions, and environmental protection in 2012. In case of any violation, Samsung would suspend transactions immediately having adopted a zero-tolerance policy [8]. Information related to product specification and environmental requirements, delivery information, and exposure to hazards, are shared between supply chain partners to facilitate environmental management in their segment of supply chain [9].

5. Stakeholders:
   Firms need to understand the needs and concerns of stakeholders, such as employees, stockholders, suppliers, neighbors, governments, peers and public interest groups, in order to incorporate their opinions to develop solutions to reduce environmental impact. For example, Samsung collects information from key stakeholders through different communication channels, including:

   – Reaching out to *customers* through customer relationship management programme, eco-friendly awareness campaign, conducting survey, and green marketing.
   – Working with *suppliers* on energy saving partnership, support for greenhouse gas inventory, and training in climate change response.
   – Training *employees* through meetings with the CEO, communication through Samsung media, internal blogs, and eco-friendly campaign.
   – Participation in *government* policy, the Environment Mark, and carbon labeling system.
   – Supporting *NGOs* through answering environmental inquiries, responding to green ranking, and producing sustainability report and information postings on its websites.
   – Communicating with *local communities* through eco-friendly campaigns with the participation of local residents, establishing a "one company, one village" sisterhood relationship, supporting children's environmental schools, and hosting CSR forums for college students.

6. Teamwork:
   Teamwork component of product stewardship is concerned with developing tightly knitted working relationships between experts, who have expertise on different aspects of the product's lifecycle. For example, while Samsung's

research and development teams search for eco-friendly materials and design products with lower environmental impacts, the manufacturing team needs to know about all the materials information and product specifications in order to carry out and monitor environmental impact of the production process without fault. Marketing team also needs to know about the product features and eco-friendly characteristics for highlight in marketing programmes. While the legal team needs to establish environmental standards that comply with national and international regulations, it also needs to ensure that all internal as well as external parties are aware of and comply with the standards.

7. Awareness:

   Organizational awareness of product stewardship is referred to as a situation where firms stay on top of industry news regarding new product changes and risk. This will allow firms to develop quick response to changes which can impact product safety such as resources, processes, science, technologies, uses, users/customers and societal and regulatory expectations.

8. Innovation:

   Innovation improves products to reduce environmental impact as well as to meet customer and societal needs. Firms are advised to actively conduct research and development for improving environmental performance in order to achieve sustainability. Through eco-innovation, Samsung introduces a wide range of products using green technologies, which decrease use of hazardous materials while saving resources and energy in production and product in-use. For example, LED televisions that are high in energy efficiency with EU energy efficiency rated A+, washing machine with the lowest energy usage in the U.S. that consumes only 90 kWh/year, hazardous materials free smartphones, and so on [8].

9. Management:

   Management of product stewardship is concerned with implementation of practices that continue product stewardship management within the firm and find new ways to improve continuously.

10. Integration:

    Lastly, integration of product stewardship is referred to action that integrate each and every function and department of the organization when practicing product stewardship. This includes every function making up the lifecycle impacts of a product. Individual consumers are also product stewards whenever they buy, use, and discard any product.

    No product is entirely environmentally safe, such that it does not require resources and consume energy to produce or to use. Yet, in order to minimize negative environmental impacts and maximize eco-friendly features of the products, product stewardship is required. The above ten components are crucial to guide firms in implementing product stewardship. In order to reduce the environmental impacts of products, involvement from the supplier, manufacturer, retailer and consumer is required.

2.1.3 Product Stewardship Example

Seiko Epson Corporation, more commonly known as Epson, a Japanese electronics company has environmental conservation as one of their social responsibilities. It implements a six-step product stewardship practice [10]:

- *Think*: Epson understands that the environmental effects of a product during use as well as throughout the product life cycle are mostly determined at the planning and design stages. Epson designs and plans with three foci: energy-saving design, resource saving, and elimination of harmful substances. Epson designs products with an aim to reduce energy consumed during use by setting an energy-conservation goal for each product. To save resources, Epson sets targets for recyclable rate of products as well as reduce consumption of resources through compact product design in terms of size and weight. Epson also develops and makes use of a database that collects all the information on banned substances by different standards.
- *Choose*: Epson strives to procure green production materials with less environmental impact. Epson has established the SEG Green Purchasing Standard for Production Material in 2003, which maps out the organization's policies to green procurement of production materials and provide standards and procedures for implementation. Epson has also developed the supplier part approval and production material approval systems to ensure all products are in compliance with product content restrictions and do not contain banned substances set by Epson's standards.
- *Create*: During the manufacturing process of its products, Epson focuses on global warming prevention, zero emissions, chemical substance management, and factory environmental risk management. The prime example is the "Epson Method", a patented technology to calculate PFC emissions, which is one of the most difficult non-CO_2 greenhouse gases to measure. This method has radically reduced Epson's level of PFC emissions. The Epson "zero emissions" program is an initiative that promotes effective use of resources, waste reduction, and extension of product lives to avoid disposal. All subsidiaries of the group worldwide are to meet material recycling targets at production sites. In addition to recycling, Epson is recently moving towards resource conservation practices [10], for example cutting energy use in basic facilities and buildings at plants, reduce energy use through production equipment, and process innovations and introducing new types of energy.
- *Deliver*: Epson has initiatives in Japan and other countries to lower CO_2 output in logistics. Epson monitors its shipping volumes, energy usage, and CO_2 emissions to reduce environmental impact of its distribution activities with a goal of reducing emissions by 1 % versus the previous year per unit of sales. To do so, for example, Epson Europe B.V. (Netherlands) introduced double-decker trucks in Europe to replace conventional trucks. These double-decker trucks increase loading efficiency, subsequently reduce the number of trucks required for the same volume of delivery.

– *Use*: Energy consumption and ease of product disassembly for recyclable components are parts of Epson's focus when designing a new product. Epson uses a variety of environmental labels to communicate to customers about their environmental performance. A "Product Information Sheet" is supplied to customers to ensure end-users can safely and properly use Epson products and know about chemical content of products.
– *Recycle and Reuse*: To treat end-of-life products, Epson established a worldwide network of collection and recycling system. Customers can recycle Epson products with ease as product recycling and reusing issues are already incorporated in the design stage.

2.1.4   Performance Impacts of Product Stewardship

A research study on the role of environmental management and the influence of environmental management capability of suppliers on firm performance and pollution reduction found that product stewardship does not directly generate desirable financial and environmental performance. Yet, product stewardship is a useful enabler for process stewardship and subsequently bringing desirable performance [11]. Selecting and evaluating alternative materials and components in product and packaging development is part of product stewardship. These practices emphasize the use of renewable, nonhazardous, and recyclable materials in product manufacturing [12]. Product stewardship prevents the use of hazardous materials in products, and subsequently reduces hazardous waste and reduces environmentally threatening risks, such as water contamination. Environmentally friendly production techniques as well as resource and energy conservation are also important for production and usage of eco-design products [13]. Product stewardship requires cross-functional participation to determine the environmental consequences of products. Specifically, it requires involvement from procurement and production through to distribution and marketing [14]. The resource-based view (RBV) of a firm advocates that organizational resources and capabilities are valuable, rare, and inimitable determine the competitiveness of firms [15]. Adapted from the resource-based view, the Natural Resource Based View (NRBV) suggested that businesses are constrained by and dependent on the conditions and resources of their natural environment to prosper and flourish [16]. A firm can make use of three interconnected strategies—pollution prevention, product stewardship, and sustainable development to achieve superior performance. According to the NRBV, manufacturers would be more capable in reducing pollution and risk of accidental polluting or hazardous substance releases.

Research studies have also found that through product and packaging eco-design, use of low-impact materials and components, and an energy efficient product distribution method, the environmental consequences of products in production and distribution can be controlled [13]. In addition, firms can benefit through promotion of their news and feature stories. The improved environmental reputation can attract and retain environmentally conscious customers [17]. For example, Hewlett-Packard and Dell Inc. are known for their global recycling programmes, and return and recycling service. These environmentally responsible initiatives have earned positive publicity and customers' compliments. Hewlett Packard and Dell Inc. h avoidance of legal penalty and bad public image in the case of environmental regulation violation [18], and cost savings from eco-design and recycling of more than 30 % of their products [19], financial performance can be improved.

However, a research study on the environmental management of Taiwanese electronics manufacturers has revealed that in some cases product stewardship brings a negative impact on the environmental and financial performance of the organizations [11]. This indicates that product stewardship is desirable but insufficient to lessen environmental damage, reduce costs in product development, and improve financial performance. In the early implementation stage of product stewardship, firms would incur costs and irrecoverable financial investments through new product design, use of recyclable containers for transportation, redesign of packaging and sourcing of ecological materials. Therefore, firms would not be able to secure the benefits of scale economy and waste reduction [20]. Similarly for pollution reduction, it is inevitable that the development of environmentally friendly product and packaging would generate waste and pollution. Therefore in the beginning of implementation, it would be difficult for firms to achieve pollution reduction.

Product stewardship requires involvement of all parties related to a product lifecycle. From the beginning of the lifecycle, the manufacturer source environmentally friendly materials and with design focusing on ease of disassembly and recycling. In the production stage, the processes with the least environmental pollution and risks would be used. Product stewardship aims to minimize waste and improve efficient use of resources through product design, packing, and material. However, prior research found limitations for cost savings and pollution control [21]. In the introduction stage of product stewardship implementation, economies of scale are difficult to achieve. Costs are likely to be incurred when developing environmentally friendly products and returning the reusable packaging to manufacturers [20].

2.1.5 Product Stewardship from the Consumer Perspective

Product stewardship concerns reducing environmental impacts with less use of hazardous and nonrenewable materials in product development. In the fashion industry, companies are similarly investing in this practice as eco-fashion. Eco-fashion consumption is a key aspect of sustainable development of fashion supply chain [22]. Chan and Wong [23] studied the consumption side of sustainable fashion supply chain. The study revealed that store-related attributes of eco-fashion positively influence consumers' eco-fashion consumption decision. The price premium level of eco-fashion also affects this decision [23]. The research was conducted on consumers in Hong Kong, to examine the relationships between product- and store-related attributes of eco-fashion and fashion consumers' eco-fashion consumption decisions.

The study found that product-related attributes were not related to eco-fashion consumption decision. This suggests that consumers are less likely to purchase eco-fashion due to physical needs for body protection and functionality, emotional needs of expressing their personality, and psychological needs of identity building. However, store-related attributes, namely customer service, store design and environment, store's ethical practices and shop convenience, are found to positively influence eco-fashion consumption and plays an important role in consumers' eco-fashion purchase decision making.

Price premium is also another important factor in regards to eco-fashion consumption decision. A high price premium would discourage eco-fashion consumption decision, as fashion consumers tend to have a tight budget or do not want eco-fashion consumption to cause any sacrifice [23]. The Hong Kong research study establishes that price premium would affect the relationship between store-related attributes and eco-fashion consumption decision [23]. This can be explained by the fact that store-related attributes are largely intangible and consumers would base their perception and purchase decision on the price premium [24]. The research also finds that price premium would negatively affect the relationship between store-related attributes and eco-fashion consumption decision [23]. This shows that consumers may have a negative perception on store-related attributes when there is usually a price premium on eco-fashion. While having the above findings in consideration, fashion companies should also devise marketing plans to promote eco-fashion consumption, subsequently facilitating the development of sustainable fashion supply chain [25].

Global fast fashion giant, H&M Hennes & Mauritz AB (H&M) illustrates the findings of the aforementioned research. H&M is known for their environmental sustainability efforts—H&M Conscious. They have seven green commitments: to provide fashion for conscious customers, choose and reward responsible partners, be ethical, be climate smart, reduce reuse recycle, use natural resources responsibly, and strengthen communities [26].

Whether fashion companies can influence fashion consumers' purchase behavior depends on their ability to fulfill fashion consumer needs [27]. H&M fulfills the need for affordable and latest fashion trends. Their sustainability efforts are widely known because of the successful on-going ethical fashion range "Conscious Collection". The collection drew such generous amount of favorable responses among the public as well as celebrities. The environmentally friendly "Exclusive Glamour" collection was seen dazzling down red-carpet events on several A-list starlets [28] in 2012. The second "Conscious Exclusive" collection of evening launched in March 2013 was also another success.

Additionally, sometimes fashion consumers are unaware of the availability of eco-fashion and are only able to acquire limited information about eco-fashion [29].

H&M satisfies this need with on-going "Conscious" promotion in-store and online. H&M also has e-tailing business in the United States and Europe in addition to their existing physical stores to improve the ease of accessing eco-fashion and the information about eco-fashion [30]. E-tailing allows easy access for potential customers to up-to-date information about eco-fashion, such as the prices and styles. Although H&M e-tailing is not available in some countries, its website still provides all the local product information and prices. This way, consumers are provided with the benefit to search and evaluate more information about eco-fashion when making a purchase decision.

Fashion consumers also have psychological needs to have symbolic feeling of advantage that links to a socially responsible lifestyle and to express their personal ethical identity through purchasing from a store that behaves in ethical ways [22]. H&M is known for adopting ethical practices, from sourcing environmentally friendly materials to garment recycling. Product stewardship starts at the beginning of the product lifecycle. H&M is the world's largest user of organic cotton. Use of organic cotton in garments is more environmentally friendly than conventional cotton. The former requires less energy and water in production; zero pesticides and 95 % of all textiles can be recycled [31]. H&M has also broadened their sustainability efforts in collaboration with the World Wildlife Foundation (WWF) to improve stewardship of their global water supply [32].

Product stewardship is not only seen in H&M production but also with old garments. H&M was the first fashion company to launch Garment Collecting, a global initiative for customers to bring any unwanted garments of any brand in any condition to H&M stores worldwide [33]. To reduce garment waste, H&M separates the old garments into three groups: rewear, reuse, and recycle. Rewear garments are sold as secondhand goods, reuse garments are no longer proper for wearing but can be transformed into other products such as cleaning cloths, or be recycled for textile fibers.

2.1.6 Summary

Product stewardship is concerned with the environmental impact of products relating to their packaging design, production, delivery, usage, and development [34]. It attends to lessen the environmental damage of products in a supply chain from materials and components sourcing, production and distribution, to disposal [35]. While the concept of product stewardship for different industries are very For example, electronics manufacturers emphasize on reducing liability and costs, therefore their product stewardship involves eco-design of products for easy dissemble of components for reuse or recycling, design of packaging to reduce materials consumption and facilitate the recycling of packaging and adoption of reusable containers for transportation [36]. For the fashion industry, on the other hand, product stewardship focuses on ethical materials sourcing, reducing energy usage in production, and recycling.

An increasing number of global fashion brands are adopting eco-fashion. H&M demonstrates extensive product stewardship practices—from the sourcing of organic cotton, reduced use of water and energy in production to garment recycling. Public relations and media on their efforts are built through on-going promotions of the "Conscious Collection" in-store and online.

Store-related attributes of eco-fashion—customer service, store design and environment, store's ethical practices and shop convenience—are important in making eco-fashion consumption decision. Fashion companies are encouraged to promote eco-fashion through satisfying consumers' needs by improving store-related attributes. Fashion companies should also adopt ethical practices and inform fashion consumers. So that the consumers' need to lead a socially responsible lifestyle would be fulfilled.

2.2 Process Stewardship

Process stewardship is a process-oriented environmental management practice [37]. It is concerned with reducing adverse environmental impact in the processes ranging from production, distribution, to end-of-life product management. It emphasizes waste reduction and contributes to environmental protection through implementing such processes as recycling reengineering polluting processes, minimizing carbon emission and so on [38]. Different from product stewardship that focuses on understanding and managing a specific product or commodity, process stewardship is about understanding and managing processes that are conducted in various phases of product development (e.g. extraction, smelting, generation of by-product) in order to minimize their potential harm to people and the environment. Mechanisms of process stewardship are useful in waste reduction and preserving the environment through managing the product flows from purchasing, storing and shipping, to asset recovery activities. This environmental management practice requires firms to implement recycling processes and develop a returned product collection infrastructure to collect, sort, and disassemble the returned products with the objective to capture reusable parts for new product manufacturing. Process stewardship would also utilize transportation capacity, reduce carbon emission, make use of cleaner transportation, and so forth. It can also benefit organizations with cost savings from for example, lowered fuel consumption.

2.2.1 Process Stewardship Example

Panasonic Corporation, a Japanese multinational electronics corporation, is committed to offer innovative and eco-friendly products through environmental management. It adheres to the ISO 26000 Core Subjects and GRI G3.1 Sustainability Reporting Guidelines to promote a consistent approach in sustainability reporting.

Panasonic's remarkable green efforts has earned the organization 4th place in the Interbrand Best Global Green Brands 2013 ranking. Its commitments include opening advanced environmentally friendly factories India, Brazil and Vietnam. Panasonic's new Econavi products—a range of air conditioners, washing machine and refrigerators—can save energy by monitoring the user's living environment and adjust the power usage to suit. The Econavi products also comply with "China Environmental Labeling Type II," the highest environmental protection product certification issued by the Chinese Ministry of Environmental Protection [39]. This new range of products not only saves energy but also is designed for reduced CO_2 emissions and using fewer materials during manufacturing.

Examples of process stewardship efforts of Panasonic include:

2.2.1.1   Chemical Substance Management

Through enforcing REACH and Restriction of Hazardous Substances Directive (RoHS) compliance measures, Panasonic strives to minimize environmental impact of their products. First, Panasonic identifies chemical substances in products. It contributes to the global goals set at the World Summit on Sustainable Development (WSSD) by disclosing publically information on all chemical substances used in its products across the supply chain. Panasonic works with suppliers to obtain information about substances contained in the parts used. Second, Panasonic evaluates the impact of the chemical substances used in the parts. It measures the amount of substances of very high concern (SVHC) that customers would be exposed to and provides information for using products with SVHC safely. Third, Panasonic discontinues use of any chemical substances that would cause any environmental risks [40].

2.2.1.2   Water Resource Conservation

To conserve water resources, Panasonic collects, treats and reuses wastewater from manufacturing processes and air conditioning systems. For example, the Technopark of Panasonic India Group recycles 100 % of the water used. After being used inside the manufacturing plant, ground water goes through wastewater treatment. The water is then reused as toilet flushing water and for lawn sprinkling. This greatly reduces fresh water use in its operations and subsequently eases the environmental risk of water shortage [41].

2.2.1.3 CO_2 Reduction

Panasonic also participates in one of the long-term global environmental targets of reducing CO_2 emissions and other green house gases by 50 % from the 2005 level by the year 2050. It has developed a unique indicator "size of contribution in reducing CO_2 emissions" to accelerate emissions reduction. Through improving the energy-saving performance of Panasonic products, CO_2 emission would be further reduced [42].

2.2.2 Performance Impacts of Process Stewardship

Process and product stewardship are both important mechanisms of environmental management [43]. A research study on the environmental management of Taiwanese electronics manufacturers revealed that the implementation of product stewardship by an electronics manufacturer is positively associated with its process stewardship [11].

Design for reuse or recycling, and application of reusable containers for transportation are useful for process stewardship. For example, electronic manufacturers can capture residual values of returned products, reuse containers that have returned products collected from the market, and reduce consumption of new inputs by utilizing reusable parts.

The research study has also found that process stewardship is positively associated with pollution reduction. Process stewardship controls transportation method, production, carbon emission, and disposal, which play important role in reducing waste and environmental impact. This proactive pollution reduction practice—which involves partners such as suppliers, customers, and logistics service providers—creates a first-mover advantage that is difficult for competitors to imitate.

The financial value of preventing pollution should not be underestimated. Process-oriented environmental practices are found to be important for both financial performance and pollution reduction. For example, environmentally conscious customers would be aware of the firm's process stewardship practices and so that they would return their end-of-life products for recycling. Process stewardship would also help reducing costs through lessening resources consumption (e.g. fuel and energy usage in operations) and capturing residual values (e.g. reusable components and materials) to lower the costs of purchasing materials and components [11]. Managers should work with suppliers who are ISO 14000 certified, conduct environmental evaluation on the second-tier suppliers, reduce environmental impact in their manufacturing processes, provide ecological proof of their outputs, and communicate about their environmental management with trading partners.

2.2.3 Summary

Process stewardship concerns using recycled or environmentally friendly materials, reducing waste, and adopting environmentally oriented technologies to lessen environmental impact in supply chain processes from production, distribution, to disposal [44]. Process stewardship is valuable for firms to reduce environmental pollution impacts and achieve financial performance. Through environmental protection practices of collecting and recycling returned products, reducing emissions and so forth, firms can attract and retain environmentally friendly consumers. Process stewardship also brings forth the benefits of product stewardship. Product stewardship efforts such as eco-design, recycling packaging, and use of nonhazardous materials for product development enables process stewardship to achieve environmental and financial performances.

The impact of product stewardship and process stewardship on environmental and financial performance is further improved if a manufacturer works with suppliers of high environmental management capability. When the environmental management capability of suppliers is high, product stewardship and process stewardship have significant positive influence on pollution reduction as the research study on the environmental management of Taiwanese electronics manufacturers has found [11]. When manufacturers work with suppliers of high environmental management capability, process stewardship also brings positive impact on return on asset, return on equity, net profit, and earnings per share. However, this relationship does not exist when working with suppliers of low environmental management capability.

2.3   Green Services

In environmental management, green practices of organizations are not confined only to the manufacturing aspect. The green services sector is also an important element to improve environmental performance through environmental management practices. The distinction between goods manufacturing and service production is irrelevant for understanding service systems [45]. Similar to product manufacturing, providing services also requires and consume resources and creates waste and pollution [46]. For example, the leisure and tourism, hotel and catering industries consumes natural resources and contributes to pollution through solid waste and water disposal [47]. Some product-oriented industries would also provide such services as logistics services, which consume fuel and emit greenhouse gas. A lifecycle study on greenhouse gas emissions has found that household consumption of services—excluding electric utilities and transportation services—is responsible for 37.6 % of total industrial greenhouse gas emissions in the United States. This amount is almost twice of what is produced through household consumption of electric utility and transportation services [48]. This case clarifies that the service industries would not have less environmental impact on the climate change than the manufacturing industries and environmental management for service industries is just as important. Jim Hanna, the director of environmental impact of international coffee chain Starbucks revealed to TheDailyGreen.com—part of Hearst Digital Media—that Starbucks is pursuing Leadership in Energy and Environmental Design (LEED) because they "have a commitment to reduce our environmental impact... [And] also realize that operating sustainably absolutely makes good business sense."[49].

2.3.1 Green Services Definition

Green service practices have been defined as the provision of services that take into account environmental sustainability [50]. The Environmental Protection Agency (EPA) of the United States has developed the Environmentally Preferable Purchasing Guidelines. According to these guidelines, environmentally preferable services can be defined as "services that have a lesser or reduced effect on human health and the environment when compared with competing products or services that serve the same purpose. This comparison may consider raw materials acquisition, product, manufacturing, packaging, distribution, reuse, operations, maintenance, or disposal of the product or service." [51].

Green services practices of service firms can be divided into *green service delivery and green service support* [52]. Green service delivery represents a firm's practices and efforts to be environmentally responsible while performing its core service businesses. It is concerned with the extent to which firms manage a mix of tangible and intangible aspects in service operations, where the combination of goods, information, and services are delivered in an environmentally friendly manner with an aim to satisfy customer expectations [53]. Ranked in the top 50 of Interbrand's Best Global Green Brands 2013, United Parcel Service of North America, Inc. (UPS) has innovative fleets and green facilities as one of their environmental sustainability efforts. In 2011, the UPS headquarters earned Leadership in Energy and Environmental Design (LEED) Gold certification from the U.S. Green Building Council and Energy Star certification from the U.S. Environmental Protection Agency. This certifies that the UPS complex uses less energy, is less expensive to operate and generates fewer greenhouse gas emissions than 75 % of similar buildings in the U.S [54]. This is an example of green service as UPS manages their complex in an environmentally friendly way where their tangible service operations consuming less energy with lesser greenhouse gas emissions. They have also recently announced one of the largest single deployments of zero-emission vehicles in the world; a fleet 100-strong that will reduce consumption of conventional motor fuel by approximately 126,000 gallons per year [55].

The peripheral or after sales services that supplement core services to enhance value to customers and service delivery are also part of service operations [56]. Green Service support is considered a peripheral service that supplements core service delivery by providing information and benefits, such as online inquiry, product collection for maintenance, and product end-of-life management, with an aim towards environmental protection. Green service support adds value to core services while reducing environmental impacts. Service support is important in addressing and meeting the expectations of customers. Additional costs may be incurred while providing service support and prior research has recommended the use of Web-based service support to enable efficient and responsive service operations [57]. For example, courier customers are often concerned with where their parcels were located and estimated arrival time. Courier services nowadays—including UPS have efficient tracking systems that allow customers to see the statuses of their parcels online. Additionally, customer enquiries could also be handled through a live chat system for UPS United States and Canada. This approach facilitates sustainable consumption of services because it requires less customer contact (e.g., less transportation) and less consumption of physical resources (e.g., paperwork) [58].

2.3.2   Green Services Example—the Hotel Industry

In the early 1990s, Scandic Hotels—Nordic Europe's biggest hotel chain—was on the verge of collapsing. The new CEO, Roland Nilsson, saved the company by introducing decentralized management and sustainable development. This new value system focused on the concept of sustainability, subsequently linked customers and employees, who were concerned for environmental responsibility. Scandic was then revived within a few years with profit and remarkable environmental sustainability efforts [59].

To stabilize the company and to grow towards the future, Nilsson believed it was essential to gear the company towards environmental sustainability, and that it would create a tighter bond with customers in a service industry like hospitality. Nilsson aimed to build customer loyalty and hence financial stability through establishing "an emotional tie between the customers and the company" based on the common desire for environmental improvement [59].

Although the restoration of profitability by Scandic relied on traditional turnaround tactics, such as cost cutting, management restructuring, and layoffs, sustainability also played a large part to their improved financial performance. Nilsson wrote in the company's 1997 annual report that their "commitment to environmental issues has also contributed to an improvement in margins, e.g., through the resource hunt…" Scandic's sustainability strategy showed the importance of green services implementation, and brought the organization visible benefits of cost savings and employee satisfaction.

The Scandic Hotels case was a prime success example of how the service industry can benefit financially from environmental protection practices. Nowadays numerous corporations in the hospitality industry have jumped on to the green services bandwagon. Hyatt Hotels Corporation is an American international company in the hotel industry. The group established Hyatt Thrive in 2011 as its corporate responsibility platform. Environmental sustainability is one of the four pillars that Hyatt Thrive focuses on. Its aim is to take focused, aggressive steps to reduce environmental impact by implementing more sustainable business practices across its operations. It recognizes the environmental impact associated with the group's hotels and therefore makes strong, focused commitment to implement more sustainable business practices.

Hyatt has been measuring the environmental performance of its managed, full-service hotels against a 2006 baseline. Its primary goals are to reduce the energy and water usage, waste sent to landfills and greenhouse gasses emission. Due to the size of their cooperation, Hyatt empowers "Green Teams" at hotels around the world to implement a wide range of esnvironmental initiatives. For example, the Green Team at Hyatt corporate headquarters in Chicago has started a competition to reduce paper consumption, which resulted in saving 2.5 million sheets of paper, amounting to a total cost savings of US$22,000 [60].

2.3.2.1 Responsible Purchasing

As an international player in the hospitality industry, Hyatt consumes numerous amounts of resources while providing their services. Therefore responsible purchasing is a large part of its global environmental protection initiatives. The amenities packaging such as shampoo and lotion bottles are made of recycled plastic, which reduces 209,000 pounds of plastic waste sent to landfills in one year. Actively encouraging guests to participate in its environmentally friendly initiatives such as providing guests with reusable cloth laundry bags and promoting towels and linin reuse. Additionally, it uses detergents that help reducing energy and water use during laundry cleaning; this alternative helps saving 168,000 therms of energy [61].

2.3.2.2 Energy Efficiency and Waste Reduction

Hyatt has invested in over US$37 million on environmental initiatives. It has completed more than 200 projects on energy efficiency improvements as of 2012. It has achieved significant reductions in the intensity values of energy use, greenhouse gas emissions, and water consumption. For example in 2012, energy intensity saw a 11 % decrease since the baseline year in 2006 and greenhouse gas emissions intensity decreased by 13 % since the baseline year as well [62].

In 2013, Grand Hyatt Beijing reduced energy use by 40 % over the past seven years. The Beijing engineering team launched several efficiency projects such as implementing central plant and lighting upgrades. These initiatives helped the hotel achieve cumulative cost savings of US$4.9 million. At the same time, the Green Team also organized hotel-wide campaigns that focused on waste reduction, switching off lights, and powering down kitchen equipment.

2.3.2.3 Consistency in Carbon Reporting

For tracking and recording greenhouse gas emissions data, Hyatt has collaborated with the World Travel & Tourism Council in the Hotel Carbon Measurement

Initiative (HCMI). The HCMI joins forces in the hotel sector in the reporting of carbon emissions, and develop a common language, which provide clarity on understanding of carbon footprint. This also helps organizations achieve their own carbon reduction goals [63]. In 2013, Hyatt integrated the HCMI framework into its in-house environmental data tracking system, EcoTrack. This allows it's reporting to be of world standards.

2.3.3   The Value of Green Service Practices

The Natural Resource-Based View (N-RBV) theory explains how green service practices create value. The resource-based view (RBV) theory credits valuable resources, capability, and business processes that are socially complex and causally ambiguous to a firm's performance improvement [15]. Business processes that exploit firm resources enable firms to utilize and expose the tangible or intangible assets to the market for performance gains, and thus becomes a source of competitive advantage [64]. An extension of the resource-based view (RBV) theory, the N-RBV recognizes the dependency of businesses on natural resources and the need for capabilities in terms of pollution prevention, product stewardship, and sustainable development to achieve lower cost, preempt competitors, and enhance future position [16].

The main premise of an N-RBV is that effective green services practices incorporate environmental protection concepts into service operations by adopting and utilizing environmentally friendly resources, technologies, and materials. And thus based on RBV, firms achieve environmental performance improvement and cost reduction from resource synergy across (intangible) organizational processes and tangible assets [64]. Taking Hyatt as an example, the new environmental sustainability practices that were implemented were intangible value creating processes that gains them a competitive edge. Making use of energy and water reducing detergents, installing energy-efficient lighting at 97 % of Hyatt hotels, and having a "Green Team" culture are all aspects that are intangible and difficult for competitors to imitate.

2.3.4   Green Services Impacts

Many service industries, such as leisure and tourism, hotel, and catering sectors involve a wide range of physical components and natural resources. A research study examining environmental management systems of service firms in Hong Kong has found that green service delivery can reduce service operation costs and environmental damage. Service firms in Hong Kong have recognized their dependency on natural resources and the need for green service capabilities have managed to lower cost and also improve environmental performance. The study shows that the adoption of appropriate green service delivery and support practices could lead to improved cost and environmental performance, as illustrated in the aforementioned hotel industry examples.

The study also demonstrates the importance of green service support practices. After sales and peripheral services contribute towards improving cost efficiency and environmental protection. The adopting green service support practices include offering online support centers, especially to those that facilitate maintenance and return collection. This supports the argument for the use of online service support for virtual interactions between service firms and customers [65], to reduce unnecessary transportation of materials and passengers [47].

For example, The Outnet.com—a fashion outlet website of Net-a-Porter the online luxury fashion retailer—offers a live chat facility for customers. For the ease of the customers, the live chat box appears on the bottom right corner of the screen during office hours. This virtual service allows customers to enquire about products and reduces the chance of purchasing a wrong item. This is an environmentally friendly way to offer help to customers and reduces unnecessary shipment and packaging materials of the products at the same time. Customer loyalty is also gained because potential shoppers would be able to get professional opinions easily before making a purchase; this greatly reduces customer disappointment by clarifying product information with them.

2.3.4.1 Environmental Management Systems

The main purpose of an EMS is to develop, implement, manage, coordinate and monitor corporate environmental activities to achieve compliance and waste reduction.[1] Compliance meaning a firm is reaching and managing the minimal legal and regulatory standards for acceptable pollution levels and would not incur sanctions. An EMS involves a formal system and database, which integrates procedures and processes for the training of personnel, monitoring, summarizing, and reporting of environmental performance information to internal and external stakeholders of the firm. This documentation is primarily internally focused on design, pollution control, waste minimization, training, reporting to top management, and so forth [66].

The International Organization for Standardization developed the ISO 14001, a certified EMS, which provides the criteria for an environmental management system. This voluntary EMS is not a set of performance standards but a framework of processes that organizations should follow to set up an effective EMS. It facilitates standardized EMS worldwide. ISO 14001 can be divided into two categories: organizational evaluation, which includes EMS, environmental auditing, and environmental performance evaluation; product and process evaluation, which

---

[1]Sayre, D., 1996. Inside ISO 14001: the competitive advantage of environmental management. St. Lucie Press, Delray, Beach, FL.

includes environmental labeling, life cycle assessment, and environmental aspects in product standards [67].

Sometimes organizations opt for uncertified EMS, which allows them more freedom to develop the most appropriate system for the organization. For example, not all of Hyatt hotels have ISO 14001 certifications, however all hotels of the organization are part of the Hyatt Thrive program making use of Hyatt Eco Track, which is an in-house central database to track environmental data such as monitoring water consumption and waste produced.

The Hong Kong research study findings also suggested that an uncertified environmental management system (EMS) strengthened the cost reduction and environmental performance impact of green services rather than a certified one. A certified EMS requires formal systems developed by a firm to adhere to international standards (e.g., ISO 14000). Whereas an uncertified EMS could have a less rigorous, but more flexible, set of environmental practices (e.g., eco-labels) [66]. A firm being ISO 14001 certified would often lead to a ceremonial behavior intended to superficially show that they are up to standards [68]. According to the NRBV, it has been argued that competitors can easily copy a formal EMS such that it is insufficient to improve operational performance. For the service industry, firms face uncertainty as their operations relies on a high level of perishability, heterogeneity, and customer involvement. Performance improvement would be difficult to achieve under a rigid system. Uncertified service firms would have the flexibility to innovate and provide environmentally friendly services without the pressure to meet audit requirements. Therefore an uncertified EMS would strengthen the relationship between green service delivery with cost and environmental performance.

The same goes for green service support. The study has found that uncertified EMS strengthened the relationship between green service support and environmental performance only. This shows that green service support is critical to both financial and environmental performance regardless of an uncertified or certified EMS is in place. As there is a high customer involvement in green service support, there is also high uncertainty that requires flexibility in managing service maintenance processes to achieve performance. Strict rules and ceremonial behavior associated with a certified EMS would not allow firms flexibility to achieve better performance with green service support. Additionally, firms such as Procter & Gamble, argued that their uncertified EMS meets or even exceeds the international standards and contributes to significant performance improvement [69]. This suggests that firms with an uncertified EMS are able to develop their own set of practices tailored for their unique operations and subsequently achieve performance improvement with green service support.

2.3.5 Summary

Green service practices can be divided into two key areas: green service delivery and green service support. Green service delivery is when a firm is environmentally responsible while delivering its service. It is an environmentally responsible practice that focuses on the use of environmentally friendly materials, conservation of materials, recycling, and energy efficiency. Green service support is a value adding peripheral service that supplements core service delivery. By offering technical support, product support, help desk support, repair services, maintenance service, recycling, training, and upgrading green service support further satisfies customer needs.

Green service delivery and green service support both contribute to cost reduction and environmental performance of service firms. Green service delivery and support would have stronger performance impact on firms with uncertified EMS in contrast to certified EMS. It has been argued that firms with a certified EMS would have rigid and strict practices in place that would actually hinder the highly versatile service industry operations and performance outcomes.

Green service delivery and support are critical to the success of environmental protection and cost reduction. Service firms need to source environmentally friendly materials, energy saving equipment, and facilities in their service delivery to reduce the use of electricity, fuel, carbon emissions, and waste disposal. Managers should be aware that green service delivery becomes more valuable to environmental and cost performance when a firm has an uncertified EMS. Green service support works with synergy to green service delivery as a peripheral service, for example if customers can access assistance online easily, customer value and satisfaction would be improved.

2.4 Logistics Management

Logistics concerns the organized movement of goods, services, and sometimes people. Logistics is the handling of the physical movement of products between one or more parties in the supply chain [70]. The Council of Supply Chain Management Professionals (CSCMP) in America—a worldwide professional association dedicated to supply chain management—defines logistics management as "part of supply chain that plans, implements, and controls the efficient, effective forward and reverses flow and storage of goods, services and related information between the point of origin and the point of consumption in order to meet customers' requirements." [71]. Logistics activities include inbound and outbound transportation management, fleet management, warehousing, inventory, logistics network designs, supply and demand planning and so forth [72]. In some cases, logistics functions can also include sourcing and procurement, packaging and assembly, and customer service.

Logistics is an energy intense sector, which has rapidly expanded due to globalization. Transportation has been recognized as one of the main sources of pollution, namely in the form of air and noise pollution [73]. Transportation was expected to increase even faster than the general growth of GNP in the industrialized world, where transportation volumes in the European Union have increased more than GNP in 15 years, since 1986 [73]. The globalization of production, procurement, and marketing has increased the 'freight transport intensity' of the economy, generating more tonne-kilometers of freight movement per unit of output. Much of energy usage has been associated with logistics, due to the rapid growth in demand for logistical services. A recent study on reducing energy consumption and emissions in the logistics sector has estimated that freight movement by truck, rail, and ship accounts for roughly a third of the total energy used by transport and around 6–7 % of total global energy consumption using statistics compiled by the IEA [74].

There are increasing concerns about negative environmental effects of logistics practices and that they are unsustainable in long run. Freight transport in the logistics sector accounts for around 8 % of energy-related CO_2 emissions worldwide [75]. The European Commission estimated that the amount of energy used to move freight might over take the energy consumed by cars and busses in the European Union by the early 2020s [76]. And that the International Energy Agency (IEA) has forecasted that—where everything remains unchanged—tonne-kms moved by trucks and ships will double between 2005 and 2050, while rail freight volumes will rise by 50 % [74]. The proportion of CO_2 emitted by logistics will rise rapidly because there are some logistics activities—such as shipping, long haul trucks and aircraft—will always rely on liquid and carbon-based fuel, subsequently requiring a large amount of fossil fuel. Therefore, mitigation of the negative environmental impacts of logistics management has drawn increasing attention.

2.4.1 Definition of Green Logistics Management

Green logistics management (GLM) aims to deploy processes that produce and distribute goods in a sustainable way, with a view to reducing waste and conserving resources in performing logistics activities. Green logistics activities include measuring the environmental impact of different distribution strategies, reducing the energy usage in logistics activities, reducing waste and managing its treatment [77].

GLM can be viewed as a management approach by which firms manage, evaluate, report, and control the environmental impacts of their actions throughout the life cycles of their products [78]. This management approach requires firms to:

1. Adopt procedures to operate, document, and report their logistics activities
2. Conduct evaluation of their performance
3. Inform and communicate with various stakeholders regarding their logistics activities
4. Comply with environmental standards

2.4.2 Drivers of GLM

Transport and logistics is a rapidly growing activity worldwide, especially when multinational organizations actively participate in global sourcing and production. With oil prices on the rise and greenhouse gas emissions concerns, optimizing logistics is one of the primary focuses for firms, in relation to both financial and environmental pressures. In 2008, BearingPoint Insight's global survey examined companies' viewpoints and practices on green supply chain management. The study collected data from over 600 professionals located in Europe, North America, and Japan. The survey found that green logistics is the link in the supply chain that attracted most interest. 81 % of the responding companies indicated that they made changes to their transport and logistics operations, out of which overhauling logistics organization to reduce the number of journeys was the most common approach, which accounts for 41 % out of all environmental measures [79]. As of 2012, in another report from 2012 done by BearingPoint interviewing more than 600 professionals worldwide, 66 % of the companies surveyed had started to green their logistics network. Out of which 37 % companies collaborate with partners to share logistics resources and 37 % and subsequently reduce environmental footprint [80]. Green logistics management enables firms to reduce pollution and energy consumption, while improve volume and cost performance. The 2012 report identified the main drivers for green logistics implementation include environmental regulations, costs reduction, and improving brand image [80].

The public is another main driver for green logistics. A research study investigating how environmental regulations, customer pressure, and economic pressure are associated with the implementations of GLM by Chinese export manufacturers finds that customer pressure is a significant factor affecting the extent of GLM implementation [81]. The findings suggest that regulatory and economic pressures are not primary considerations of manufacturers, whereas the GLM implementation is stimulated more by legitimacy and relationship continuation with customers. Similarly, a survey done by Eyefortransport—a business intelligence provider in the transport, logistics and supply chain industry—found the same general message in 2007, where the top two key drivers for initiating green logistics are to improve public relations and improve customer relations [82].

The research study based on Chinese manufacturers finds that economic pressure is not significant enough as a primary driver for firms to implement GLM. As the aforementioned surveys done by BearingPoint and Eyefortransport have mentioned, the primary driver is on satisfying customer requirements. Manufacturers implement GLM when they experience a higher level of the ecological modernization forces, in terms of environmental regulations, customer pressure, and economic pressure. The performance outcomes are positively associated with the extent of GLM implementation. Firms pursue GLM can reap environmental, operational, and financial benefits [81]. The surveys done by Insight (2008) and Eyefortransport (2007) also find that in addition to customer pressures, companies implement GLM for reducing logistics costs and gaining return on investment [82].

2.4.3 Performance Impacts of GLM

GLM can be reflected in the adoption of various environmental management practices such as reducing carbon dioxide emissions in distribution, remanufacturing, reuse, recycling, extending the product life cycle, and capturing value from used products [83]. By collecting and analyzing returned products, firms can identify problems related to product use patterns. This will allow firms to identify and develop breakthroughs for product improvement and new product development. This indirect market feedback provides valuable input to firms for improvement in product design, sourcing decisions, forward and reverse logistics planning, and new product development [84]. By implementing GLM, firms can reap the benefits of reducing the environmental burden of developing, distributing, and disposing of products, while improving financial, and operational performance.

The research study investigating how environmental regulations, customer pressure, and economic pressure are associated with the implementations of GLM by Chinese export manufacturers finds that GLM mediates the relationships between the ecological modernization forces—as characterized by customer, economic, and regulatory pressures—and their environmental, financial, and operational performance [81]. GLM is highly valuable to the ecological modernization of firms, in Chinese firms in particular. With the environmental problems in China worsening day by day to a point of eroding its GDP growth [85], the country calls for a balance between its economic development and environmental conditions [86]. Firms need to efficiently use natural resources such as energy, fuel, raw materials, and water, as well as cost and environmentally effectively manufacture and distribute products with economic value. Through pursuing GLM, firms can improve their environmental, financial and operational performance.

In 2010, McKinsey & Company conducted a survey examining how companies manage sustainability activities and how they assess and communicate these engagements of 1946 executives representing a wide range of industries and regions. The findings reveal that more than 50 % of executives consider sustainability "very" or "extremely" important in regards to new-product development, reputation building, and overall corporate strategy [87]. According to BearingPoint Insight's global survey on green supply chain management, firms recognize the benefits of green logistics in terms of improved brand image, customer satisfaction, differentiation from competitors, costs reduction, and so forth [79].

2.4.4 Example of Green Logistics Management

Kraft Foods Group Inc., an American grocery manufacturing and processing multinational firm recognized the potential environmental and cost impact of logistics operations and has aimed at improving its distribution network efficiency since 2005. In 2009, Kraft announced that it has saved more than 50 million truck miles since 2005 in its transportation sustainability efforts. As quoted in Kraft Foods' press release, Steve Yucknut, Kraft Foods' vice president of sustainability stated, "We are finding ways to drive fewer miles, reduce inventory piles, and eliminate idling trucks. We are collaborating with customers and suppliers. And we are using a number of high-tech innovations for our trucks and warehouses to reduce energy and CO_2 emissions." [88].

2.4.4.1 Increased Rail Use

Rail networks are less polluting however, not many logistics operations opt for this transport, as it is relatively less flexible. Kraft Foods started with "riding the waves (and rails) to environmental efficiency" as stated in its press release. It has greened their logistics through shipping via water channels and railways. Instead of shipping wheat to a Toledo, Ohio flourmill by truck, Kraft Foods saved more than a million miles (1.6 million km) by shipping via waterways. Replacing 10,000 truck shipments and reduced 2000 tons of CO_2 emissions [88]. Kraft Foods also sends products to customers by train instead of trucks in the United Kingdom and Austria, saving more than 40,000 miles and 150,000 miles respectively. Kraft Foods has gained a first mover advantage as the recent BearingPoint study reveals that the share of rail use is low in product transportation, representing only 5 %. However, Deutsche Bahn—European railway operator—expects an increase in demand for green rail operations, as customers will prefer green products more and more. Additionally, transporting goods by rail is a more competitive option for organizations in the long run, as oil prices are very volatile. French rail operator SNCF also revealed that it would increase operations in cargo transportation [80].

2.4.4.2 Fuel-Efficient Fleet Optimization

Saving fuel and energy is another key aspect for greener logistics. Inefficiencies in freight operations result in empty trucks most of the time. Kraft Foods addressed this by establishing a single hub in Bratislava, Slovakia, and has modernized its transportation network in Europe reducing 20 % fewer trips between its European plants and distribution centers. Additionally, the firm has utilized a special software to optimize truck loads that helps to reduce more than a million miles [89]. These direct store delivery vehicles and electronic refrigeration technology use up to 30 % less fuel a conventional truck. Optimizing freight operations to help saving energy consumption is not only applicable to large organizations like Kraft Foods, but also smaller firms and even self-employed individuals. For example, under the Guangdong Green Freight Demonstration Project—a World Bank initiative—has helped optimizing truck freight. The Guangdong government has helped Lin'an—a logistics brokerage information center that connects distributors with truck drivers —expand its online business [90]. This has greatly reduced the amount of empty trucks inefficiently wasting time and fuel in the province.

2.4.4.3 Technology Investments and Collaboration

Kraft Foods has started working with the U.S. Environmental Protection Agency—a market-driven partnership aimed at helping businesses move goods in the most efficient way possible—to work out the best practices and latest technologies best for the firm. Kraft Foods participated in the EPA's SmartWay Transport Partnership where the firm adopted a no-idle engine policy at its shipping locations, piloted a hybrid frozen delivery truck program and increased its use of intermodal transport. Philippe Lambotte, senior vice president of customer service and logistics at Kraft Foods told Consumer Goods Technology that "full trucks that never stop... We believe we can move from approximately 12 percent of our volume moving on cooperatively routed lanes to a possible 30 percent," [91].

Additionally Kraft Foods has collaborated with Oracle Transport management and rolled out the Project MOST (Management Optimized Sustainable Transportation) initiative. It has worked with Oracle consultants to develop the software that enhances performances and is functional for the company. The software can analyze Kraft Foods' complex network of locations, measures truck movements, analyzes routes and devises new tracks to eliminate empty miles. Project MOST has cut more than 500,000 miles in 2009. "Overall, this transportation management project has allowed us to create unique capabilities for our business units and customers, allowing Kraft Foods to balance cost and service in a sustainable manner. We believe this will raise the bar of our performance levels for years to come," Lambotte commented [91]. The green logistics efforts continued, in 2010, Kraft Foods has increased 36 % of sustainable sourcing, cut 45 million pounds of packaging, and eliminated 20 million kilometers of travel miles [92].

Apart from technology advancements that require a large amount of monetary investment, smaller firms can start with partnering with other logistics organizations. This has also become a trend in recent years, the BearingPoint report has revealed that more than 37 % companies interviewed are working with partners to share logistics resources in order to reduce environmental footprint. The report also forecasts a significant increase in logistics collaboration [80]. Procter & Gamble (P&G) has set a target to reduce road kilometers travelled by trucks by 30 % in the United Kingdom by 2015, and it has already achieved the target in 2013. The company revealed that the key is collaboration, working with retail partners and logistics suppliers; it also saw a trend in collaboration with firms in the same industry as well [93].

2.4.5 Summary

Globalization and multinational organizations made a jump in the transportation and logistics operations in the past decade. Without doubt, there are increasing concerns on the various environmental detrimental impacts of these operations. For example, energy consumption and greenhouse gas emissions draws primary interests for companies with truck shipments. Companies with green logistics management (GLM) aim to implement processes and practices that produce and distribute goods in an environmental way, with a focus to reduce waste and conserve resources while performing logistics activities. Measuring environmental impact of distribution strategies, reduction in energy usage and reducing waste in logistics activities are all part of green logistics practices.

The main drivers for organizations to implement GLM are namely customer pressure, public relations, energy saving and subsequently cost saving. GLM allow firms to reduce pollution and energy usage while still achieving the same or improved volume and cost performance. Customer pressure is a significant factor affecting the extent of GLM implementation. Research findings suggest that customer pressure is a primary consideration rather than regulatory and economic pressures. GLM implementation is stimulated more by legitimacy and relationship continuation with customers.

GLM implementation is positively correlate with environmental, financial, and operational performance. As GLM requires that enterprises adopt various environmental management practices such as reducing carbon dioxide emissions in distribution, remanufacturing, reuse, recycling, extending the product life cycle, and capturing value from used products. Through collecting and analyzing returned products, firms also acquire valuable indirect feedback for product problems and product development. GLM also mediates the relationships between the ecological modernization forces—namely customer, economic, and regulatory pressures—and their environmental, financial, and operational performance. The current environmental situation worldwide calls for a balance between economic and environmental performance. Firm can achieve this balance by efficiently using resources such as energy, fuel, raw materials, and water, as well as effectively manufacture and distribute products that create customer value. Through pursuing GLM, firms can improve their environmental, financial and operational performance.

Organizations and business managers should be aware of the environmental impact of logistics activities. Implementing GLM is a crucial part of the greening of supply chains. Firms should also work with logistics providers to communicate and share a mutual agreement on the importance of environmental protection. Implementing GLM requires participation of the whole organization. Top management commitment and staff training as well as human resources participation such as campaigns to raise staff awareness can support the adoption of green logistics.

2.5 Green Logistics Management

2.5.1 Green Logistics Management

Green logistics management (GLM) is a management approach by which firms manage, evaluate, report, and control the environmental impacts of their logistics operations throughout the product life-cycle (Sroufe 2003). Firms implementing

GLM often: (i) adopt procedures to operate, document, and report logistics operations, (ii) evaluate of the logistics performance, (iii) communicate with various stakeholders regarding their logistics activities, and (iv) comply with environmental standards and regulations [94]. GLM are involved with reducing pollution in distribution, remanufacturing, reuse, recycling, capturing value from used products, and extending the product life cycle. GLM enables firms to produce and distribute goods in a sustainable way so as to reduce waste and conserve resources in logistics performance [77].

Lai and Wong [1] identified that there are four attributes of GLM:

- *Procedure-based practices*:
  Procedure-based practices are the practices to perform GLM based on corporate structure and reporting systems in firms. These practices serve as a communication tool between organizational functions about their responsibility in GLM.
- *Evaluation-based practices*:
  Evaluation-based practices reflect organizational ability to evaluate, monitor, and improve performance on a periodic basis. They aim to offer formal documents and reports to managers, reducing equivocality on the success of GLM.
- *Partner-based practices*:
  Partner-based practices reflect organizational ability to coordinate with internal staff members and external partners in information sharing and communication on the development of GLM. A firm not only should provide staff communication and training on GLM but also should enhance partnership with suppliers and customers so as to competently mitigate the environmental impact arising from product flows.
- *General environmental management practices*:
  General environmental management practices require visibility in product development and compliance with global environmental standards along the logistics chain. Firms maintain records in satisfying environmental standards with a formal reporting system for stakeholder access, such as publishing corporate social responsibility reports [95].

2.5.2 Driver of Green Logistics Management

The drivers of GLM can be broadly classified into customer pressures, economic incentives, and regulatory requirements.

- Economic Incentive
  GLM is an important environmental practice that satisfies the growing organizational quest for productivity, while reducing pollution and resources consumption. It helps address the environmental issues incur in a product life-cycle.

For instance, distribution of products in an environmentally friendly manner would require guidelines to ensure product recovery can be achieved efficiently with compliance requirements and partner collaboration. Firms encounter two types of economic pressures in the global market, namely dematerialization and decoupling [95]. Dematerialization is referred to lower consumption of natural resources for each unit of output, while decoupling is concerned with lower dependency on the input of natural resources for continuous productivity growth. In view of economic pressures, GLM is a viable approach to resolve cost challenge and respond to the environmental request. Examples of economic incentive include governmental subsidies and material costs savings.

Rapid industrialization in recent years increases public's concern about the polluted air and water, energy shortage, and deforestation, posing challenges to firms in conserving natural resources in their operations (Economy and Liebertha [85]). Resource scarcity and rising raw material prices (e.g. crude oil, copper, nickel, steel, and resin) lead to an increase of material costs, which in turns raises economic pressure to firms to take environmental initiatives, such as using reprocessed or scrap materials in new product development. For example, Kraft Foods Group Inc., an American grocery manufacturing and processing conglomerate, engaged in making its distribution network more efficient and used a combination of alternative transportation and technology to reduce its logistics footprint. Motivated by saving the expenses on energy use, Kraft has been finding ways to drive fewer miles, reduce inventory piles, and eliminate idling trucks. In order to reduce emission, Kraft ships wheat to a Toledo, Ohio, flour mill by ship rather than by truck, eliminating the equivalent of 10,000 truck shipments and roughly 2000 tons of carbon dioxide emissions. Meanwhile, hybrid technologies are being used for store delivery vehicles to reduce fuel consumption and carbon emission (GreenBiz staff 2009).

- Customer Pressure

The rise of corporate social responsibility has laid foundation for transparency and accountability in sustainability performance of firms [85]. The environmental disclosures serve as a channel for firms to communicate and share information about their environmental performance, which helps shape public perception and environmental reputation of firms. A green image enhances the reputation of firms, which helps gain acceptance in the global market, especially for those export-orientating firms that are required to comply with environmental requirements and market expectations in sales to foreign customers. These requirements and expectations are imposed by customers to fulfill their own environment-related obligations, as well as to demonstrate their environmental responsibility [85]. The customer-based environmental requirements vary from ISO 14000 certification to retrieval of reusable components or products from their countries.

GLM implementation can generate spillover effects to nurture customer preference for related products so as to avoid environmental incidents and the consequential legal costs and fines [1]. GLM implementation not only facilitates and promotes product return and recycling services, but also cultivates a firm's positive image through feature news to attract and retain environmentally conscious customers [17]. Customers often boycott companies and their products if the company is found with violation of environmental laws [96]. Many anecdotes suggest that sourcing from environmentally irresponsible manufacturers can damage the reputation of customer firms (e.g. retailers). Therefore, there is a rising trend for firms seeking suppliers who are ISO 14000 certificated, and monitoring suppliers performance through environmental auditing, in order to satisfy the environmental expectations of their customers. The implementation of GLM may signal the congruence of the "going green" expectation of the international community.

For example, Maersk Line, the shipping company responsible for around 1 percent of global carbon dioxide emissions, has saved US$764 million on fuel over 2013 by cutting its carbon dioxide output by 12 percent. Maersk's eco-performance improvement is largely driven by the increase in its customer demand. Large customers representing 19 % of Maersk's business have requested tailored sustainability information as part of the service requirements. To comply with its customer requests, Maersk moves goods with a smaller environmental impact for its customers, delivering their sustainability promises (BusinessGreen 2014).

• Environmental Regulatory Requirement
Environmental regulatory pressure is concerned with regulations enacted by local and oversea regulatory authorities to control environmental impacts of organizational activities ranging from production, transportation, to disposal [1]. In many cases, these pressures are mandatory for firms to comply with due to legal requirements. Firms are required to comply with the environmental policies, such as WEEE and REACH [97], when exporting their products to the EU market. The products must be free from hazardous substances in accordance with Restriction of Hazardous Substances Directive (RoHS) and Registration, Evaluation, Authorization and Restriction of Chemicals (REACH). Also, the manufacturers and suppliers are required to undertake the responsibility of collection, treatment, and recycling of end-of-life products. These regulatory pressures reflect the environmental concerns by the local community and regulatory bodies as negative externalities caused by imports [98].

RoHS, was imposed in February 2003 by the European Union and its directive took effect in July 2006. Toshiba, a Japanese multinational engineering and electronics conglomerate corporation, took actions immediately to address the EU environmental directives and unveiled its first RoHS-compatible notebook, the Tecra S3, for the business channel in October 2005. Toshiba also introduced the first RoHS-compatible notebook with the Satellite A55-S1064 at its retailer Wal-Mart in December 2005 (GreenBiz Staff 2007). RoHS is closely linked with the Waste Electrical and Electronic Equipment Directive (WEEE), which sets collection, recycling and recovery goals for electrical products and is part of a legislative initiative to resolve the problem of toxic e-waste. Since 2004,

Panasonic, Thomson, and Victor Company of Japan (JVC) cooperated to develop a recycling scheme for the electronics and electrical equipment industries to comply with the requirements of WEEE. Their cooperation agreement include (1) establishing a recycling scheme in EU, (2) supervising the entire recycling operations, and (3) inviting other manufacturers and recyclers to join the scheme (GreenBiz staff 2004).

Based on survey data from 128 Chinese export manufacturers, Lai et al. [94] found that customer pressure is significantly and positively associated with the extent of their GLM implementation while environmental regulations and economic incentive are less likely to affect the manufacturers' pursuit of GLM. Lai and Wong [1] further suggested that the positive effect of customer pressures on GLM implementation is enhanced by environmental regulatory pressure.

2.5.3 Summary

GLM helps minimize the environmental impact of forward and reverse flows of products, information and services between the point of origin and the point of consumption. GLM can be driven by three incentives—economic pressure, customer pressure, and environmental regulation. Governmental subsidies and material costs savings provide economic incentives for firms' green logistics operations. In addition, firms need to comply with environmental regulatory requirements and address customer pressure if their products are to compete in global markets.

2.6 Extended Producer Responsibility

Extended producer responsibility (EPR) originated from Europe as a policy concept aimed at extending producers' responsibility for their products to the post- consumption stage of product life with the presumption that manufacturers have the capability to manage the end-of-life products to reduce environmental impacts [99]. The Product Stewardship Institute defines EPR as a type of product stewardship that includes, at a minimum, producers are responsible for post-consumer management of their products as well as packaging [100]. EPR in some countries, such as the European Union, South Korea, and China, is mandatory in such context as WEEE and ELV Directives, which require producers to be responsible in managing product-life-cycle from production to post-consumption stages of products and their packaging. Governments promote EPR for two major reasons. First, the policy allows local governments to be relieved from financial burdens of waste management. Second, the policy encourages reduction of primary resources consumption, urging manufacturers to utilize alternative materials and commence product design innovations for reducing disposal and waste in production activities [101].

The notion of EPR is largely different from green supply chain management, green purchasing, and corporate environmental management. Yet, the implementation of EPR requires supply chain efforts, and the support of these activities is crucial to its success. Green supply chain management focuses on the inter-organizational efforts in managing the supply chain processes to reduce adverse environmental impact from materials purchase, production, to distribution of finished products [81]. EPR, on the other hand, extends green supply chain management to manage take-back, recycling, and final disposal of products. As EPR manages residual values of returned products, its implementation is highly related to green purchasing that takes account of organizational sourcing decision with a focus on reducing use of environmentally unsustainable materials by developing purchasing policy, defining environmental objectives, and monitoring environmental performance of suppliers [102]. Green purchasing is involved in sourcing environmentally sustainable raw materials to facilitate recycling of returned products. It is also responsible to identify reusable components and parts of returned products for new product development to reduce purchase of virgin materials. As such, green purchasing supports EPR implementation by ensuring use of environmentally sustainable raw materials and making use of residuals retrieved from returned goods. Moreover, EPR is different from the concept of corporate environmental management that is confined to organizational efforts and practices to reduce their adverse environmental impact through product and process stewardship with an emphasis on reducing liability and costs [103]. Nonetheless, EPR requires the support of these efforts and practices to ensure environmentally friendly product design for disassembly and recycling, and to enable and facilitate production of new product and recycling of returned goods to be done with low level of environmental impact.

EPR emphasizes on the principle of waste prevention by manufacturers with practices such as recycling, reprocessing, and reusing the components or materials with residual values. Encouraging manufacturers to use sustainable materials and design sustainable products for recycling, and consequently reducing disposal, waste, and consumption of resources is an important goal of EPR. Many industries such as the automobile [104] and electronics [105] industries have established standards as reference for manufacturers to develop corresponding solutions for mitigating disposal and waste of reusable materials or components caused by their industrial activities. EPR practices can be a feasible way for manufacturers to seek more sustainable forms of development by improving their overall eco-efficiency in product development.

2.6.1 Approaches of EPR

EPR can be applied voluntarily by organizations or mandated by governments. EPR is implemented differently in different countries and organizations, sharing the common theme of having the producer bearing larger responsibility for product end-of-life management. The following are some of the most common approaches.

2.6.2   Product Take-Back Programs

Take-back requirement is a major element of an EPR policy, mandating manufacturers to collect and treat end-of-life products. This policy is highly beneficial for manufactures to incorporate environmental consideration at the product design stage to facilitate their subsequent take-back programs. This product stewardship emphasis improves and advances the treatment of returned products [106] by disassembled parts inspection, reusable parts separation, recycling, reprocessing, and reusing the parts with residual value in the product take-back process [107]. This collection of EPR practices enhances producer ability to competently satisfy both the local and international requirements on environmental protection.

Some firms offer voluntary product take-back program to be environmentally responsible. For example, Apple Inc., an American multinational electronics company, offers voluntary take-back programs in different parts of the world. Apple Inc. reuse and recycling program varies somewhat differently across countries, but runs along the same lines, such that customers can return their old Apple Inc. electronics for free and discount will be offered to these customers for purchasing new Apple Inc. products. Taking the Apple Inc. U.S. Recycling Program as an example, customers can trade-in their iPhone, iPad, Mac, or PC to get an Apple Inc. Store Gift Card of the estimate market value of the returned product [108]. The implementation of product take-back program not only encourages customers to return their old products for recycling, but also promotes customers to continue to use the company's products.

A mandatory take-back program is when a government imposes regulations to mandate manufacturers or retailers to take-back their end-of-use products. For example, the Federal Ministry for the Environment, Nature Conservation and Nuclear Safety of the Federal Republic in Germany enforced the End-of-life Vehicle Act, which is applicable to all vehicles sold in the Germany market after 1 July 2001. The regulation requires manufacturers and importers to set up nationwide collection system or commission third parties to collect their end-of-life vehicles. Along with this regulation, a recycling target was set, manufacturers, importers, and the third-party waste management firms have to meet the recycling rate goal of 80 % from 2006 onwards [109].

2.6.3   Advanced Recycling Fee (ARF)

An advanced recycling fee (ARF) is a fee added on top of product sales to cover the cost of recycling in California, U.S. Customers are charged with an extra fee when purchasing an electronic product. This fee would be used for the recycling of the product, allowing responsible disposal of electronic products. In California, ARF is imposed in the form of the Electronic Waste Recycling Fee under the Electronic Waste Recycling Act of 2003. Retailers collect the fee on the electronic devices from consumers for end-of-life products management. The fee varies depending on the size of the devices, ranging from USD$3 to USD$5 [110].

2.6.4 Voluntary Industry Practices

In some countries, some firms of the same industry group together to develop a set of voluntary EPR guidelines, which provides a set of instructions for recycling and waste diversion tailored for the particular industry. For example, the Carpet America Recovery Effort (CARE) in the U.S. is founded by representatives of the industry, and it develops market solutions for the recycling and reuse of post-consumer carpet. The organization's mission "is to advance market-based solutions that increase landfill diversion and recycling of post-consumer carpet, encourage design for recyclability and meet meaningful goals as approved by the CARE Board of Directors." [111]. In 2012, CARE members have recycled 294 million pounds of carpet and diverted 351 million pounds of carpet from landfill, which was an increase of 17 and 5 %, respectively, in comparison to 2011. The organization has diverted over 2.6 billion pounds of post-consumer carpet from landfills since it's foundation in 2002 [112].

2.6.5 Example of EPR

Sony Electronics Inc. has based its Green Management 2015 mid-term environmental targets on the idea of EPR, where it "strives to achieve an environmentally conscious recycling system and effective operation for collection and recycling of end-of-life products." [113]. The following are Sony's various EPR efforts.

2.6.5.1 Take-Back and Trade-in Programs

Sony was one of the first companies to launch a nationwide electronics-recycling program in the U.S. in 2007 [114]. The Sony Take Back Recycling Program encourages consumers to recycle and dispose electronics equipment responsibly. Consumers can drop off Sony products at collection centers free of charge. In the beginning, Sony started a partnership with Waste Management, a waste management company in the U.S., with 75 collection centers. Currently, Sony extends its partnership with other waste management company and works in collaboration with different waste administration and recycling companies in U.S. with 417 collection centers [115].

Sony also developed a website dedicated for trade-in and recycling electronic products in the U.S. Consumers can gain website credits by trading in electronic products, even non-Sony ones, and can search for recycling centers to drop off devices with no trade in value. Website credits gained can be redeemed when purchasing Sony products. As of March 2013, Sony has cumulatively collected 280 million pounds of electronics equipment scrap for recycling, reducing disposal of electronics to landfill and using of virgin materials significantly.

2.6.5.2   Mandatory Compliance

In some countries, Sony is required to comply with specific EPR regulations. For example, the Regulation for the Administration of the Collection and Disposal of Waste Electric and Electronic Products in China, more commonly known as the China WEEE [116], requires Sony to be responsible in treating waste electronic and electronic equipment products. This regulation aims to conserve natural resources and develop a circular economy, subsequently achieving environmental protection. The regulation establishes a system for end-of-life electronics collection and disposal of WEEE products with a state-established and state-administered WEEE Disposal Fund supported by payments from manufacturers and producers [117]. Sony makes regular contributions to the fund and treats its waste electronic in compliance to this regulation.

2.6.5.3   Voluntary Industry Practices

The Japan Home Appliance Recycling Law (revised in 2009) covers LCD/Plasma televisions, refrigerators, washing machines, air conditioners and clothes dryers. This law requires consumers to be responsible for the fees associated with disposing of these home appliances for retailers to take back and for manufacturers to recycle. To comply with this regulation, Sony partners with five other manufacturers in Japan to establish a nationwide cooperative recycling network to collect and recycle their returned products. This regulation also requires manufacturers to achieve recycling rate targets of minimum 55 %. In 2012, Sony has achieved a recycling rate of CRT televisions and flat screen televisions of 81 and 89 %, respectively [118].

2.6.6   Impacts of EPR

EPR encourages manufacturers to treat their products at the end-of-life to protect the environment and reduce costs incurred from the developing landfill. The aim of this policy is to reduce waste generation at the source, promote environmentally friendly product design, and facilitate achieving the public goals of reduction, recycling, and reuse in materials management.

2.6.7 Benefits

Success in addressing environmental concerns can create business value and gain market acceptance. Recent studies find that practicing EPR can improve the performance of enterprises with less environmental damage [106]. Manufacturers can gain from recycling used products in many ways. For example, manufacturers can extract and collect rare earth materials for recycling and reduce consumption of virgin materials for manufacturing new units. Manufacturers would also be in a better position to negotiate for exchange terms in business with international partners that emphasize waste prevention and disposal in their local market [119].

The product stewardship emphasis in EPR policy accentuates environmental burden reduction with less use of hazardous and nonrenewable materials for product manufacturing. EPR can attract and retain environmentally conscious customers by offering take-back and recycling of end-of-life products, while producing new products with recycled materials. By doing so, manufacturers can gain an environmentally friendly corporate image, which is a valuable intangible asset difficult for competitors to imitate. Such a positive image also helps gain customers' confidence in the environmental impact of the products and increases their patronization of products [120].

Modular design in EPR facilitates ease of product components disassembly and recycling, reducing the use of virgin materials. Modular product design also increases flexibility of manufacturing systems in the supply chain [121]. Product modularity allows manufacturers to assemble a product from a set of smaller components, which can function collectively as a whole [122] and form different products in different variations of combinations. Such product stewardship also brings benefits to manufacturers in pollution reduction and financial gains due to cost saving through waste reduction [11]. For example, Levi Strauss, an American jeans and casual wear manufacturer, has established a returns-processing system, known as R.I.S.E. (returns, irregulars, samples, exit strategy). The system allows Levi Strauss to shorten the time needed to process returned products from weeks to days by sorting the returned products and shipping them to different locations for reprocessing or resell [123]. The system helps Levi Strauss increases turnover of the returned products to save costs and generate incomes.

Modular designs encourages recycling and reuse of product components, where reusable parts are valuable for recovery of assets and cost containment concerning returned products that reduces materials acquisition costs and inventory requirements. For instance, Gap Inc, an American multinational clothing retailer, has a "solid waste and recycling program" that uses less corrugated cardboard and utilizes recyclable materials for its containers. This program reduces cardboard waste by 5700 tons and saves Gap US$20 million per year. EPR practices, including recycling, reuse, reprocessing, and recovery, are also helpful for generating new source of revenue by capturing reusable components to be sold at after-market [124]. IKEA is an example of capturing residual values of returned products from the market. IKEA uses its returned and damaged products as spare parts and restores them to a saleable condition to be sold at a discount. This practice helps capture the residual values of products, and lessen consumption of virgin materials by utilizing reusable parts recovered.

2.6.8 Moderating Role of Customer Integration in EPR Performance

Customer integration is concerned with the participation of customers in product return process and their attention and efforts made to facilitate the manufacturers' EPR practices. For manufacturers to meet EPR goals, integrating with customers is vital. EPR emphasizes on managing post-consumer products, however it also implicitly assumes the requirement for a degree of customer involvement that they would fulfill their responsibility voluntarily by returning the end-of-life products to the manufactures [125]. Manufacturers are required to develop a system to facilitate the take-back and collection of products returned from customers in local as well as other overseas to implement EPR policy.

The contingency theory views a firm as an open system, where the firm's business environment affects its performance [126]. Organizational processes are endogenous, which are mainly decided and controlled by firms, whereas the business environment is exogenous that firms have relatively less control over [127]. According to this theory, EPR practice performance is not only dependent on a firm's ability but also the level of integrating with customers in support of the EPR implementation. The EPR practices-performance relationship depends on how well customers take up their responsibility to retrieve usable products from the markets and take part in product return programs. A high level of customer integration can help improving corporate image and organization positioning in the marketplace, subsequently strengthening the relationship between EPR practices and market performance.

A recent research study examining the market and financial performance of EPR practices by manufacturing exporters in China finds that integration of customers on EPR practices to recycle, reprocess, and reuse their products can improve market acceptance of firms [128]. Customer integration in EPR practices allows manufacturers to develop an environmentally responsible corporate image, and increases opportunities of accessing international markets. The study results show that manufacturers seeking to establish market position with their EPR practices should strengthen customer integration with their environmental initiatives. For example, proactively inviting customers to be part of their environmental audits and product return program to encourage more support in their EPR practices.

Customer integration, however, causes uncertainty to the implementation of EPR practices from the financial perspective. Uncertainty is increased in the firms' environmental efforts because the amount of products returned and delivery time is highly dependent on customer willingness to cooperate [129]. This increased uncertainty would incur costs to manufacturers because of additional organizational efforts, such as managing inventory and incurring repair costs. Customer integration may lead to excess or shortage of returned products, which makes inventory management difficult for manufacturers [130]. Subsequently, the financial performance of EPR practices is compromised when customer integration is high.

Prior research study on Chinese manufacturers also finds that firms with EPR practices implemented with lower level of customer participation achieve better financial performance [128]. This indicates that firms need to develop a mechanism to coordinate with customers for EPR implementation, to reduce costs in handling returned products and streamline cross-firm processes [95]. Managers should implement an efficient system for customer returns and inventory to control the cost of coordination, so that financial performances would not be jeopardized. Partnering with other manufacturers to establish a large returns system or with recycling firms would improve the effectiveness in managing EPR practices. For example, Apple partners with PowerOn Service Inc., one of the industry leaders in the responsible reuse and recycling of post-consumer information technologies and electronics, for its reuse and recycle program.

2.6.9 Summary

EPR is defined as a type of product stewardship that requires at a minimum the producer to be responsible for their product and that responsibility extends to the end-of-life processes. In some countries, such as South Korea and China, EPR is mandated by regulations. Governments promote EPR mainly to decrease waste management financial burdens and to provide incentives for reducing resources consumption. EPR encourages manufacturers to make use of sustainable materials and to design products suitable for recycling, subsequently reducing waste through recycling, reprocessing, and reusing components.

There are various approaches of EPR, the most common types are namely take-back programs, advanced recycling fee, and voluntary industry practices. Take-back programs can be voluntary or mandatory. Take-back programs can help manufacturers to determine product problems and use this information in product improvement. It also allows them to capture residual values and to reuse working components, which helps to save materials costs. Advanced recycling fee (ARF) is a fee charging customers on top of product sales to cover the cost of the post-consumer life of the purchased product. Voluntary industry practices are undertaken when different firms from the same industry come together to develop a set of voluntary EPR principles for the industry.

Performance benefits of EPR include improved market performance. Where firms can gain market by producing products manufactured with recycled materials with less environmental damage. Additionally, residual materials extracted from returned products can help saving the cost for virgin materials consumption. Modular design in EPR can improve ease of product components disassembly for recycling and reuse, enabling financial gains due to cost savings.

Customer integration has a moderating role on the performance of EPR practices. Customer integration can benefit market performance of a firm implementing EPR. Customer integration helps firms to develop an environmentally responsible corporate image, and increase opportunities of accessing international markets. Managers should consider reinforcing customer awareness on their responsibility, and to remind them of the environmental impacts of their return of usable products.

References

1. Lai K-H, Wong CWY (2012) Green logistics management and performance: some empirical evidence from Chinese manufacturing exporters. Omega 40(3):267–282
2. Snir EM (2001) Liability as a catalyst for product stewardship. Prod Oper Manage 10 (2):190–206
3. Reinhardt FL (1998) Environmental product differentiation: Implications for corporate strategy. Calif Manag Rev 40(4):43–73
4. Lamming R, Hampson J (1996) The environment as a supply chain management issue. Br J Manage 7(Special Issue): S45–S62
5. Samsung (2013) Our sustainable development at samsung electronics. Available from: http:// www.samsung.com/us/aboutsamsung/citizenship/oursustainabilityreports.html. Cited 19 Dec 2013
6. Product Stewardship Institute (2013) I. What is product stewardship? Available from: http:// www.productstewardship.us/. Cited 19 Dec 2013
7. Adams G (2010) 10 Principles of responsible product stewardship, in GreenBiz.com. 2010
8. Samsung (2013) Sustainability Report 2013
9. Lau HCW et al (2002) Monitoring the supply of products in a supply chain environment: a fuzzy neural approach. Exp Syst 19(4):235–243
10. Corp SE (2013) Environmental product development lifecycle. 2013. Available from: http:// global.epson.com/SR/environment/lifecycle/index.html. Cited 9 Dec 2013
11. Wong CWY et al (2012) Green operations and the moderating role of environmental management capability of suppliers on manufacturing firm performance. Int J Prod Econ 140 (1):283–294
12. Drumwright ME (1994) Social responsible organizational buying: environmental concern as a noneconomic buying criterion. J Mark 58(2):1–19
13. van Hamel C, Cramer J (2002) Barriers and stimuli for ecodesign in SMEs. J Clean Prod 10 (5):439–453
14. Carter CR, Kale R, Grimm CM (2000) Environmental purchasing and firm performance: an empirical investigation. Transp Res Part E 23(3):219–228
15. Barney J (1991) Firm resources and sustained competitive advantage. J Manag 17(1):99–120
16. Hart SL (1995) A natural-resource-based view of the firm. Acad Manag Rev 20(4):986–1014
17. Schuler DA, Cording M (2006) A corporate social performance—corporate financial performance behavioral model for consumers. Acad Manag Rev 31(3):540–558
18. Buttel FH (2000) Ecological modernization as social theory. Geoforum 31(1):57–65
19. Hindo B, Arndt M (2006) Everything old is new again. Bus Week 3999(1):65–70
20. Russo MV, Fouts PA (1997) A resource-based perspective on corporate environmental performance and profitability. Acad Manag J 40(3):534–559
21. Lewis H, Gertsakis J (2001) Design and environment. Greenleaf Publishing, Sheffield
22. Niinimäki K (2010) Eco-clothing, consumer identity and ideology. Sustain Dev 18(3): 150–162

23. Chan T-Y, Wong CWY (2012) The consumption side of sustainable fashion supply chain: understanding fashion consumer eco-fashion consumption decision. J Fashion Mark Manage 16(2):193–215
24. Dodds WB, Monroe KB (1985) The effect of brand and price information on subjective product evaluations. In: Hirschman EC, Holbrook MB (eds) Advances in consumer research. Association for consumer research vol 12. pp 85–90
25. Gurau C, Ranchhod A (2005) International green marketing: a comparative study of British and Romanian firms. Int Mark Rev 22(5):547–561
26. AB, H.M.H.M. H&M Conscious. 2013. Available from: http://about.hm.com/en/About/Sustainability/HMConscious/Aboutconscious.html#cm-menu. Cited 9 Dec 2013
27. Solomon MR, Rabolt NJ (2004) Consumer behavior: In Fashion. Prentice Hall, Englewood Cliffs, NJ
28. Karmali S. H&M Launches new conscious partywear collection. 2013. Available from: http://www.vogue.co.uk/news/2013/03/19/h-and-m-launches-conscious-exclusive-partywear-collection. Cited 9 Dec 2013
29. Joergens C (2006) Ethical fashion: myth or future trend? J Fashion Mark Manage 10(3):360–371
30. Chiang K-P, Dholakia RR (2003) Factors driving consumer intention to shop online: an empirical investigation. J Consum Psychol 13(1–2):177–183
31. AB, H.M.H.M., H&M Conscious Actions Highlights 2012. 2013
32. Interbrand. Best Global Brands 2013: H&M. 2013. Available from: http://www.interbrand.com/en/best-global-brands/2013/HM. Cited 9 Dec 2013
33. AB, H.M.H.M. Garment Collecting. 2013 Available from: http://about.hm.com/en/About/Sustainability/Commitments/Reduce-Reuse-Recycle/Garment-Collecting.html. Cited 9 Dec 2013
34. Chen C (2001) Design for the environment: a quality-based model for green product development. Manage Sci 47(2):250–263
35. Porter M, van der Linde C (1995) Green and competitive: ending the stalemate. Harvard Bus Rev 73:120–134
36. González-Benito J, González-Benito O (2005) Environmental proactivity and business performance: an empirical analysis. Omega 33(1):1–15
37. Christmann P (2000) Effects of "best practices" of environmental management on cost advantage: the role of complementary assets. Acad Manag J 43(4):663–680
38. Jabbour CJC (2008) In the eye of the storm: exploring the introduction of environmental issues in the production function in Brazillian companies. Int J Prod Res. doi:10.1080/00107540802425401
39. Interbrand. Best Global Green Brands 2013: Panasonic. 2013. Available from: http://www.interbrand.com/en/best-global-brands/Best-Global-Green-Brands/2013/Panasonic. Cited 9 Dec 2013
40. Corporation P (2013) Environment: chemical substance management. Available from: http://panasonic.net/sustainability/en/eco/chemical/. Cited 11 Dec 2013
41. Corporation, P (2013) Environment: water resource conservation. Available from: http://panasonic.net/sustainability/en/eco/water. Cited 11 Dec 2013
42. Corporation P (2013) Environment: CO_2 reduction. Available from: http://panasonic.net/sustainability/en/eco/co2/. Cited 11 Dec 2013
43. Preuss L (2011) In dirty chains? Purchasing and greener manufacturing. J Bus Ethics 34(3/4):345–359
44. Gilley KM et al (2000) Corporate environmental initiatives and anticipated firm performance: the differential effects of process-driven versus product-driven greening initiatives. J Manag 26(6):1199–1216
45. Vargoa SL, Magliob PP, Akakaa MA (2008) On value and value co-creation: a service systems and service logic perspective. Eur Manag J 26(3):145–152
46. Burja V, Burja C (2009) Increasing service quality through environmental performance management. Ann Univ Apulensis Series Econ 11(2):938–944

47. Ellger C, Scheiner J (1997) After industrial society: service society as clean society? Environmental consequences of increasing service interaction. Serv Ind J 17(4):564–579
48. Shu S (2006) Are services better for climate change? Environ Sci Technol 40(21):6555–6560
49. Canter, L. McDonald's, Starbucks & Arbys among chains for LEED. 2010. Available from: http://www.fooddigital.com/sectors/hotels-and-restaurants/mcdonald-s-starbucks-arbys-among-chains-leed. Cited 3 Jan 2014
50. Foster ST, Sampson SE, Dunn SC (2000) The impact of customer contact on environmental initiatives for service firms. Int J Oper Prod Manage 20(2):187–203
51. Agency, U.S.E.P. Environmentally Preferable Purchasing (EPP). 2013. Available from: http://www.epa.gov/epp/pubs/about/faq.htm. Cited 3 Dec 2013
52. Wong CWY, Wong CY, Boon-itt S (2013) Green service practices: performance implication and the role of environmental management systems. Service Sci 5(1):69–84
53. Fitzsimmons JA, Fitzsimmons MJ (2004) Service management, 4th edn. McGraw Hill Irwin, Boston, MA
54. United Parcel Service of America, I. Green Facilities. 2013. Available from: http://www.sustainability.ups.com/Environment/Innovative+Fleets+and+Facilities/Green+Facilities. Cited 13 Dec 2013
55. Interbrand. Best Global Green Brands 2013: UPS. 2013. Available from: http://www.interbrand.com/en/best-global-brands/Best-Global-Green-Brands/2013/UPS. Cited 17 Dec 2013
56. Roth AV, Menor LJ (2003) Insights into service operations management: a research agenda. Prod Oper Manage 12(2):145–164
57. Sampson SE, Froehle CM (2006) Foundations and implications of a proposed unified services theory. Prod Oper Manage 15(2):329–343
58. Jaruwachirathanakul B, Fink D (2005) Internet banking adoption strategies for a developing country: the case of Thailand. Internet Res 15(3):295–311
59. Goodman A (2000) Implementing sustainability in service operations at Scandic Hotels. Interfaces 30(3):202–214
60. Corp H (2013) Green teams. Available from: http://thrive.hyatt.com/greenTeams.html. Cited 31 Dec 2013
61. Corp., H. Global Initiatives. 2013. Available from: http://thrive.hyatt.com/globalInitiatives.html. Cited 31 Dec 2013
62. Corp., H. (2012) Corporate Responsibility Report 2012
63. Council WTT (2013) Hotel Carbon Measurement Initiative. Available from: http://www.wttc.org/activities/environment/hotel-carbon-measurement-initiative/. Cited 17 Dec 2013
64. Ray G, Barney JB, Muhanna WA (2004) Capabilities, business processes, and competitive advantage: choosing the dependent variable in empirical tests of resource-based view. Strateg Manag J 25(1):23–37
65. Smith RJ, Eroglu C (2009) Assessing consumer attitudes toward off-side customer service contact methods. Int J Logistics Manage 20(2):261–277
66. Melnyk SA, Sroufe RP, Calantone R (2003) Assessing the impact of environmental management systems on corporate and environmental performance. J Oper Manage 21(3):329–351
67. Tibor T, Feldman I (1995) ISO 14000: a guide to the new environmental management standards, p 230
68. Boiral O (2007) Corporate greening through ISO14001: a rational myth? Organ Sci 18(1):127–162
69. Rondinelli D, Vastag G (2000) Panacea, common sense, or just a label? The value of ISO 14001 environmental management systems. Eur Manag J 18(5):499–510
70. Wood DF et al (2002) International logistics. AMACOM
71. Professionals, C.o.S.C.M. CSCMP supply chain management. Available from: http://cscmp.org/about-us/supply-chain-management-definitions. Cited 17 Jan 2014

72. Wei H-L, Wong CWY, Lai K-H (2012) Linking inter-organizational trust with logistics information integration and partner cooperation under environmental uncertainty. Int J Prod Econ 139(2):642–653
73. Aronsson H, Brodin MH (2006) The environmental impact of changing logistics structures. Int J Logistics Manage 17(3):394–415
74. Inderwildi O, King SD (2012) Energy, transport, & the environment: addressing the sustainable mobility paradigm. Springer, Berlin, p 721
75. Ribeiro SK, Kobayashi S (2007) IPCC Fourth Assessment Report: Climate Change 2007. In: Intergovernmental panel on climate change
76. Transport ECD-G.f.E.a. (2003) European energy and transport trends to 2030. National Technical University of Athens, E3M-Lab, Greece
77. Sbihi A, Eglese RW (2007) Combinatorial optimization and green logistics. 4OR Quart J Oper Res 5(2):99–116
78. Sroufe R (2003) Effects of environmental management systems on environmental management practices and operations. Prod Oper Manage 12(3):416–431
79. Inc., B., 2008 Supply Chain Monitor "How mature is the Green Supply Chain?". 2008
80. Inc., B., Green Supply Chain: from awareness to action. 2011
81. Sarkis J, Zhu QH, Lai KH (2011) An organizational theoretic review of green supply chain management literature. Int J Prod Econ 130(1):1–15
82. McKinnon A et al (2010) Green logistics: improving the environmental sustainability of logistics. Kogan Page Publishers
83. Linton JD, Jayaraman V (2005) A conceptual framework for product life extension. Int J Prod Res 43(9):1807–1829
84. Jayaraman V, Luo Y (2007) Creating competitive advantages through new value creation: a reverse logistics perspective. Acad Manag Perspect 21(2):56–73
85. Economy E, Liebertha K (2007) Scorched earth: will environmental risks in China overwhelm its opportunities? Havard Bus Rev 85(6):88–96
86. Zhu Q et al (2011) Evaluating green supply chain management among Chinese manufacturers from the ecological modernization perspective. Trans Res Part E Logistics Transp Rev 47(6):808–821
87. Company M (2010) How companies manage sustainability. McKinsey Global Survey results
88. Foods K (2005) Kraft foods eliminated more than 50 Million truck miles since 2005 through focus on transportation sustainability efforts. Available from: http://www.prnewswire.com/news-releases/kraft-foods-eliminated-more-than-50-million-truck-miles-since-2005-through-focus-on-transportation-sustainability-efforts-70454622.html. Cited 16 Jan 2014
89. Staff G (2009) Kraft trims millions of shipping miles through smarter logistics. Available from: http://www.greenbiz.com/news/2009/11/20/kraft-trims-millions-shipping-miles-through-smarter-logistics
90. Staff TWB (2013) Improving fuel efficiency and reducing emissions of trucks in China. 2013. Available from: http://www.worldbank.org/en/news/feature/2013/09/13/improving-fuel-efficiency-and-reducing-emissions-of-trucks-in-china. Cited 6 Jan 2014
91. Ackerman A (2009) Kraft foods drives change. Available from: http://consumergoods.edgl.com/case-studies/Kraft-Foods-Drives-Change49354. Cited 6 Jan 2014
92. Foods K (2011) Our Progress in 2011
93. Waters H (2013) Supply chains of the future: sustainable logistics and profitability go together. Available from: http://www.theguardian.com/sustainable-business/supply-chain-future-sustainable-logistics-profit
94. Lai KH, Wong CWY, Cheng TCE (2012) Ecological modernisation of chinese export manufacturing via green logistics management and its regional implications. Technol Forecast Soc 79:766–770
95. Zhu QH, Geng Y, Lai K-H (2010) Circular economy practices among Chinese manufacturers varying in environmental-oriented supply chain cooperation and the performance implications. J Environ Manage 91(6):1324–1331

96. Innes R, Sam AG (2008) Voluntary pollution reductions and the enforcement of environmental law: an empirical study of the 33/50 Program. J Law Econ 51:271–296

97. Christmann P, Taylor G (2006) Firm self-regulation through international certifiable standards: determinants of symbolic versus sustantive implementation. J Int Bus Stud 37:863–878

98. Atasu A, Van Wassenhove LN, Sarvary M (2009) Efficient take-back legislation. Prod Oper Manag 18:243–259

99. OECD (2001) Extended producer responsibility: a guidance manual for governments. OECD, Paris

100. Inc., P.S.I. Home. 2013. Available from: http://www.productstewardship.us/. Cited 9 Dec 2013

101. Link S, Naveh E (2006) Standardization and discretion: does the environmental standard ISO14001 lead to performance benefits. IEEE Trans Eng Manage 53(3):508–519

102. Chen C-C (2005) Incorporating green purchasing into the frame of ISO 14000. J Clean Prod 13(4):927–933

103. Nicol S (2007) Policy options to reduce consumer waste to zero: comparing product stewardship and extended producer responsibility for refrigerator waste. Waste Manage Res 25(3):227–233

104. Milanez B, Buhrs T (2009) Extended producer responsibiltiy in Brazil: the case of tyre waste. J Clean Prod 17(5):608–615

105. Khetriwal DS, Kraeuchi P, Widmer R (2009) Producer responsibility for e-waste management: key issues for consideration—Learning from the Swiss experience. J Environ Manage 90(2):153–165

106. Subramanian R, Gupta S, Talbot B (2009) Product design and supply chain coordination under extended producer responsibility. Prod Oper Manage 18(3):259–277

107. Chung C-J, Wee H-M (2008) Green-component life-cycle value on design and reverse manufacturing in semi-closed supply chain. Int J Prod Econ 113(2):528–545

108. Inc., A (2014) Apple recycling program. Available from: http://www.apple.com/recycling/gift-card/. Cited 14 Jan 2014

109. Federal Ministry for the Environment, N.C., Building and Nuclear Safety (2010) End-of-life vehicles. Available from: http://www.bmub.bund.de/en/topics/water-waste-soil/waste-management/types-of-waste-waste-flows/end-of-life-vehicles/. Cited 14 Jan 2014

110. (CalRecycle), C.D.o.R.R.a.R. Electronic product management electronic waste recycling fee. Available from: http://www.calrecycle.ca.gov/electronics/act2003/retailer/fee/. Cited 14 Jan 2014

111. Effort CAR (2014) About CARE. Available from: http://www.carpetrecovery.org/about.php. Cited January 2014

112. Effot CAR (2012) 2012 Annual Report. Available from: http://www.carpetrecovery.org/pdf/annual_report/2012_CARE-annual-rpt.pdf. Cited January 2014

113. Corporation S (2014) Sony's policy on recycling products. Available from: http://www.sony.net/SonyInfo/csr_report/environment/recycle/policy/. Cited 14 Jan 2014]

114. America, S.C.o. (2013) Sony establishes first nationwide electronics recycling program with waste management's recycle America. Available from: http://www.sony.com/SCA/company-news/press-releases/sony-electronics/2007/sony-establishes-first-nationwide-electronics-recy.shtml. Cited 14 Jan 2014

115. Corporation S (2014) CSR Reporting. Available from: http://www.sony.net/SonyInfo/csr_report/environment/recycle/america/index.html. Cited 14 Jan 2014

116. 廢棄電器電子產品回收處理管理條例 (Regulations for the Administration of the Collection and Disposal of Waste Electric and Electronic Products), in 551. 2009: China

117. McElwee C (2011) Environmental Law in China: mitigating risk and ensuring complianceironmental law in China: Mitigating Risk and Ensuring Compliance. Oxford University Press, Oxford, p 348

118. Corporation S (2014) Recycling of television sets. Available from: http://www.sony.net/SonyInfo/csr_report/environment/recycle/japan/index1.html. Cited 14 Jan 2014

119. Ginsberg JM, Bloom PN (2004) Choosing the right green marketing strategy. MIT Sloan Manage Rev 46(1):79–84
120. Fombrun CJ (1996) Reputation: realizing value from the corporate image. Harvard Business School Press, Boston
121. Sanchez R, Mahoney JT (1996) Modularity, flexibility, and knowledge management in product and organization design. Strateg Manage J 17(Winter Special Issue):63–76
122. Agarwal R, Karahanna E (2000) Time flies when you're having fun: cognitive absorption and beliefs about information technology usage. MIS Q 24(4):665–694
123. Anonymous Levi Strauss gets a leg up on reverse logistics. IIE Solutions, 2001
124. Barnett ML, Salomon RM (2006) Beyond dichotomy: the curvilinear relationship between social responsibility and financial performance. Strateg Manag J 27(11):1101–1122
125. Forslind KH (2005) Implementing extended producer responsibility: the case of Sweden's car scrapping scheme. J Clean Prod 13(4):619–629
126. Van de Ven AH, Drazin R (1985) The concept of fit in contingency theory. In: Cummings LL, Staw BM (eds) Research in organizational behavior, vol 7. JAI Press, New York, pp 333–365
127. Astley WG, Van de Ven AH (1983) Central perspectives and debates in organization theory. Adm Sci Q 28(2):245–273
128. Lai KH, Wong CWY, Lun VYH (2014) The role of customer integration in extended producer responsibility: a study of Chinese export manufacturers. Int J Prod Econ 147:284–293
129. Guide VDRJ, Jayaraman V, Linton JD (2003) Building contingency planning for closed-loop supply chains with product recovery. J Oper Manage 21:259–279
130. Guide Jr., VDR et al (2000) Supply chain management for recoverable manufacturing practices. Interfaces, 2000. 30(3): 125–142

Chapter 3
Collaborative Environmental Management

Abstract This chapter focuses on the collaboration of environmental management between suppliers and customers, with the selected topics ranging from environmental information integration, supplier operational adaptation, supplier/customer environmental integration, and consumer preferences. Environmental information integration (EII) is the information sharing infrastructure that supports environmental information exchange and coordination, which covers internal EII, supplier EII, and customer EII. EII is important factor for organizations to increase flexibility and promote their environmental adaptability to the changing environmental demands and requirements. Establishing a long-term supplier relationship and cooperatively investing in development and innovation with supply chain partners can be beneficial to enhance an organization's operational flexibility, which in turns improves the organization's environmental and economic performance. On the other hand, customer's involvement in implementing environmental practices participation facilitates the firm's extended producer responsibility practices and asset recovery processes. Since consumers care about the environment protection which enables their descendants to have a better life, sustainable consumption such as energy conservation, resource savings, food waste reduction, and the use of recyclable products have been gaining growing public awareness. Sustainable consumption behavior may prompt organization to rethink about environmental management practices to fulfill their final consumers' requirements.

3.1 Environmental Information Integration

Environmental protection is an important organizational issue that can affect long-term development of enterprises [1]. Environmental management is no longer the responsibility of an individual firm but involves its supply chain partners. Increasingly, organizations realize the importance of supply chain collaboration in implementing environmental management practices, such as closed-loop supply

© The Author(s) 2016
C.W.Y. Wong et al., *Environmental Management*,
SpringerBriefs in Applied Sciences and Technology,
DOI 10.1007/978-3-319-23681-0_3

chains and reverse logistics [2]. To do so, such industry leaders as Nike, Hewlett-Packard and S.C. Johnson, share information and communicate with supply chain partners in support of their efforts to eliminate the use of toxic materials, reduce energy consumption, and avoid waste production in their supply chain processes.

However, as supply chain coordination can be complex, the organization of environmental management involving multiple parties to create synergy and bring upon positive results can be challenging [3]. This suggests the important role of devising an information sharing mechanism beyond an individual firm to support environmental management in a supply chain. For example, PepsiCo shares information and integrates with its suppliers to successfully reduce their greenhouse gas emissions in the supply chain as a whole [3]. PepsiCo started working with Carbon Disclosure Project (CDP) Supply Chain in 2007 to discuss the importance of emissions management with suppliers. Through conducting 3-day training courses on the assessment tool at the suppliers' manufacturing facilities, PepsiCo educated and informed suppliers about the potential cost reduction opportunities.

3.1.1 Definition of Environmental Information Integration

The information sharing infrastructure that supports environmental information exchange and coordination across business functions and partner firms is considered the environmental information integration (EII) [4]. EII is different from traditional supply chain information integration that is limited to supporting such supply chain activities as shipment coordination, order fulfillment, and so forth [5]. It is also different from vertical integration that is concerned with corporate ownership. EII represents organizational electronic connectivity that allows firms to capture and disseminate information to coordinate environmental management practices in a supply chain, ranging from eco-product design, asset recovery, components disassembly and recycling, to reuse, with the aim of reducing the environmental impact of products throughout their life cycle [6]. For example, customers, such as distributors and retailers, can participate in asset recovery activities by returning products for inspection and separation. While collecting these returned products can reduce disposal and capture the residual values of products, sharing of information with such upstream partners as component suppliers and manufacturers, related to the condition and amount of retrieved components and materials enables opportunities for product and process improvement, and plan for inventory.

To prevent environmental damages of the product processing and use, it is important for firms to generate and disseminate environmental information. For a firm, there are four types of environmental information to be share: organizational information for use internally (i.e. environmental goal documentation), organizational information for use externally (i.e. environmental reports for stakeholders and authorities), product information for use internally (i.e. performance evaluation indicators for products), and product information for use externally (i.e. eco-labels and environmental claims) [7]. EII could help the focal firm to build an environmental-friendly corporate image and deliver a signal to business partners that its operation are environmentally sustainable and social responsible. EII with suppliers and customers can increase the supply chain coordination so as to lower the risk of information asymmetry [8].

3.1.2 Dimensions of Environmental Information Integration

EII requires involvement from different participants of the supply chain including the focal manufacturer and its suppliers and customers. Hence, the formation of EII requires different dimensions of integration in a supply chain. Prior research highlighted three types of environmental relationships between a focal firm and its suppliers and customers: (i) environmental requirements, (ii) sharing environmental information, and (iii) collaboration for improving environmental aspects of products and processes [9]. Environmental requirements are referring to the company policies such as purchasing requirements, employee training, and ISO 14000 certification. These requirements guide the environmental management practices of firms by providing specifications and principles of making decisions about such functions as materials sourcing, supplier selection, and so forth. On the other hand, sharing of environmental information is concerned with sharing such information as emission data, environmental regulation and policy updates, and environmental practices performance results. The sharing of environmental information allows partner firms to assess their supply chain environmental impact. Lastly, collaboration for environmental management is referring to the co-development of environmental processes and pollution mitigation practices between functions and supply chain partners.

A firm is influenced by external forces and conditioned by internal processes [10]. In accordance with the classic view of organizations, EII can be categorized into internal and external EII, which is in line with the conceptualization of supply chain integration [11]. Supply chain integration refers to the extent to which a focal firm collaborates with supply chain partners and facilitates data flow internally and externally across its organizational processes. Therefore, from the perspective of supply chain management, EII can be divided into three dimensions: internal EII, customer EII and supplier EII.

Internal EII supports the coordination of internal processes to achieve corporate environmental goals. It represents the extent a firm shapes its communication and information sharing infrastructure across internal organizational functions to facilitate organizational efforts in environmental management [12]. For example, Inditex, a Spanish multinational clothing company with labels such as Zara and Massimo Dutti, has established the "Code of Conduct and Responsible Practices" for employees. This code dictates that "the operations of the Inditex Group shall be developed under an ethical and responsible perspective; … all the activities of Inditex shall be carried out in the manner that most respects the environment, promoting biodiversity preservation and sustainable management of natural resources." [13]. Inditex also listed in its "Strategic Environmental Plan 2011–2015" that all employees are trained to raise their awareness regarding global environmental issues and their specific implications at workplace. The initiatives include providing awareness-raising content and environmental information in corporate magazine (100,000 copies in twelve languages were distributed), and environmental training for store staff, etc. [14]. Through such internal processes as establishing an environmental code of conduct for employees to follow, training employees on environmental issues, as well as raising their awareness, helps the facilitation of organizational efforts in environmental management.

External EII includes customer EII and supplier EII, which take account of the differences between upstream and downstream supply chains. Customer EII is concerned with the information sharing framework of which firms share environmental information with customers (e.g. new eco-product introduction), while acquiring information of new customer requirements and expectations (e.g. new environmental standards and regulations). Customer EII helps identifying business and environmental management opportunities [15], while informing and educating customers about alternative products and practices. Such EII can therefore be useful in engaging customers to participate in environmental protection (e.g. returning end-of-life products). For example, Inditex is committed to the publics 'right to know' and therefore publicly discloses chemical substances used in producing its products [16]. The Inditex 'Right to Know policy' allow customers to access information about the uses and discharges of chemicals based on the reported quantities of releases of hazardous chemicals to the environment. Inditex also records and monitor feedback that customers give to staff on the shop floor, noting what sells and what does not. This information is then reported to the in-house designers at the headquarters who would quickly develop new designs, and new clothes can quickly be manufactured based on these feedback [17]. Such quick and specifically tailored response to market demands also means that there would be less chance of overstock. As a result, costs of attending to excess stock can be avoided. At the same time the risk of having scandals regarding inappropriate treatment of unsold clothing can be lowered. Case in point, the slashed unsold H&M and Walmart clothing scandal in 2010 [18]. Brand new H&M and Walmart clothing were found destroyed and discarded on the streets in New York. H&M representatives later made a public apology for such an inappropriate act that affect people and the environment.

Supplier EII is concerned with the information sharing infrastructure that enables information sharing between a focal firm and its suppliers. This facilitates upstream supply chain operations on such aspects as materials selection and treatment of manufacturing by-products. Product and process improvement can be achieved by sharing information related to their material wastage, alternative materials, manufacturing technologies, and so forth. Supplier EII facilitates the environmental objectives alignment of supply chain partners while collaboratively engaging supply chain partners in environmental management practices to reduce their adverse environmental impact. Inditex, for example, requires suppliers "to comply with environmental standards established by Inditex including, if applicable, the necessary measures to reduce and compensate such impact in order to apply said standards." [19].

Inditex provides guidelines to suppliers to ensure all materials used in its products comply with its standards. It also has a master plan for water management in the supply chain to ensure water is properly treated in order to reach "zero discharge" goal by 2020. The plan "is used to guide the efforts of Inditex and its suppliers in the sustainable use of this vital resource. One of the elements Inditex considers fundamental in the work carried out in tandem with its suppliers is the commitment to Zero Discharge of Hazardous Chemicals (ZDHC)." [13]. Inditex recognizes that its commitment to achieve "zero discharge" of hazardous substances by 2020 although goes beyond legislation, it believes that such supply chain collaborative efforts will lead towards a cleaner environment [13].

3.1.3   EII Examples

Exchanging and sharing environmental information with supply chain partners can bring upon many benefits. Prior research examined how customer and supplier relations affect environmental performance suggests that a reciprocal learning process between customers and suppliers occurs as firms exchange information to set and meet environmental requirements [20]. Close relationship and communication between supply chain partners allow new knowledge to be created and shared for environmental management. For example, firms can gain insight for new product development based on information collected from customer EII. When environmental management practices such as internal EII and supplier EII practices are adopted, continued improvement in environmental performance is more likely to occur due to capacity of firms to acquire new information for environmental management practices improvement [21].

3.1.3.1   Sustainable Apparel Coalition

The Sustainable Apparel Coalition (SAC) is a non-profit organization in the United States. It seeks to address the apparel and footwear industry's social and environmental challenges. Members include apparel or footwear brands, manufacturers, retailers, industry affiliates and trade association. Led by influential executives, such as the chief merchandising officer of Walmart and founder of Patagonia, SAC commenced its operations in 2009 with a mission to "creating a single approach for measuring sustainability in the apparel sector will do much more than accelerate meaningful social and environmental change." [22]. The SAC envisions "an apparel and footwear industry that produces no unnecessary environmental harm and has a positive impact on the people and communities associated with its activities." [23]. To bring together collective efforts as an industry, the organization developed the Higg Index in 2012 to assess and score sustainability throughout a product's life cycle in the apparel and footwear industry.

Information sharing with suppliers often occurs in one direction, such that a focal firm demands environmental information input from its suppliers as buying firms traditionally have the upper hand on picking which suppliers and manufacturers to work with. However, such an unidimensional flow of information can hinder environmental performance outcomes. This is because, first, the mandate sharing of information and environmental audits of suppliers accentuates the lack of trust in the supply chain [24]. Second, manufacturing complexity nowadays requires each distinct party plays a specific role in environmental improvement. Without sharing environmental information upstream would mean that suppliers would not be able to tackle environmental problems efficiently and effectively due to lack of market information, such as customer expectations. Similarly, the Higg Index 1.0 only facilitates unidimensional communication which requires manufacturers to share their scores to their customers, such as fashion brands and retailers. In line with the concept of EII, Higg Index 2.0 released in 2013 requires brands and retailers to share their scores with manufacturers as well. Environmental information sharing allows each party of a supply chain to take responsibility for its own environmental improvement [24].

3.1.3.2 Carbon Disclosure Project (CDP) Supply Chain Program

The CDP Supply Chain Program launched in 2008 to promote information sharing and innovation between CDP Supply Chain members. The program "enables organizations to implement successful supplier engagement strategies, reduce supply chain emissions, control water impact, and manage risk in a changing climate." [25]. It provides a wealth of information on how firms work with their suppliers and customers to reduce environmental impacts. As of 2014, the annual CDP supply chain information request generated the largest response with 2868 companies, supplying 64 supply chain program member companies. The information disclosed includes carbon emissions and approach to climate risk management [26].

Through participation in the program, sharing environmental information has helped large corporations to reap significant benefits. Working with supply chain partners effectively drives reductions in the environmental and social impact. With information integration, firms can achieve constructive communication with suppliers and listen to the needs of customers to identify environmentally sustainable opportunities that add monetary value and deliver environmental performances. For example, Walmart encourages supplier to become more sustainable and makes working partners aware of environmental improvements. It has requested MeadWestvaco Corp. (MWV), an American packaging company, to develop a more environmentally efficient package for its retail pharmaceutical adherence business.

The new paperboard-based packaging system, known as ShellPak® Renew, is about 7–80 % more greenhouse gas efficient to produce and its compact design reduces transportation costs and emissions. MWV's customers will realize greenhouse gas emission savings with this new packaging of more than 12,000 tonnes annually [26].

3.1.4 Performance Impacts of EII

Information technology (IT) applications can improve information sharing and processing capabilities of partner firms by replacing paper-based communications [27]. Prior research with data collected from 188 trading firms in Hong Kong finds that supplier operational adaptation mediates the association between IT-enabled transport logistics and cost performance [28]. IT-enabled transport logistics means the use of IT to facilitate transport logistics activities, such as cargo tracking, warehousing, and shipment notice forwarding, helping product movement in the supply chain. Involving information technology in transport logistics allows timely information sharing among partner firms [5]. Moreover, the value of sharing information for supply chain partners is widely recognized with benefits of cost reduction [29].

IT coordinates business process between partner firms, which establishes cooperation and builds trust in economic exchange and subsequently improving their relationships [30]. IT-enabled transport logistics can provide a transactional mechanism to improve mutual understanding, facilitating the development of supplier operational adaptation in the product flow process [31]. Supplier operational adaptation can serve as a relational mechanism that offers flexibility for firms to manage unforeseen operation needs, such as fulfilling orders with a short lead time and allowing flexibility in order changing [27]. The sharing of information across firms can lead to a better inter-organizational relationship in such forms as cooperation and operational adaptation [32].

An environmental uncertainty study suggests that firms hoping to reduce uncertainty in their task environment would seek collaborative relationships with suppliers [33]. IT-enabled transport logistics is helpful for uncertainties reduction in logistical coordination by increasing the availability of information requiring responsive organizational actions. Forming a collaborative relationship, partners are willing to adapt which reduces coordination efforts in transport logistics activities, subsequently improving cost performance. The Hong Kong research provides further evidence that when supplier operational adaptation plays an important contributor role under environmental uncertainty, cost performance can be improved by using IT to support transport logistics activities [28].

Collaborative planning, forecasting, and replenishment (CPFR) is the most recent prolific management initiative that provides supply chain collaboration and visibility. By following CPFR, companies can dramatically improve supply chain effectiveness with demand planning, synchronized production scheduling, logistics planning, and new product design [34].

EII is also positively associated with firms' profit-related performance. As firms disclose their information with supply chain partners, they will attain better responsiveness and detect the inefficiency to improve their services [29]. Through information exchange, firms can make business decision faster and with less uncertainty [35].

One key goal of sharing environmental information with supply chain partners is to improve a firm's environmental-related performance. EII is helpful to enhance the ability of firms to develop eco-friendly products and processes. Supplier EII enables suppliers to understand eco-design for reduction, recycling, and reuse in materials management throughout the life-cycle of products [8]. Firms need to engage in supplier EII for eco-design to facilitate the take-back activities and innovation in eco-friendly products. For example, Esprit, a Hong Kong manufacturer of clothing, regularly had workshops with suppliers sharing the company's efforts in sourcing sustainable products to find new solutions. Customer EII can also help mitigate the environmental damages caused by the products, in particular during the product return process [36]. Customer EII is instrumental in obtaining customer support and improving corporate environmental images [8]. If customers are willing to participate in the recycling and final disposal, firms will achieve their environmental goals more easily.

3.1.4.1 Customer EII

Wong [37] examines how EII contributes to environmental management capabilities in terms of corporate environmental innovativeness and adaptability. Through collecting data from 230 firms in Hong Kong participating in voluntary environmental management schemes, the study finds that customer EII engenders both corporate environmental innovativeness and adaptability. Corporate environmental innovativeness is regarded as the development and adoption of new management practices to improve a firm's environmental protection efficiency and effectiveness [38]. Whereas corporate environmental adaptability is referred to a firm's responsiveness and flexibility in coping with new environmental requirements while sustaining economic growth [39]. Customer EII facilitates effective information sharing and communication with customers. It improves the ability of firms to innovate new environmental management products and practices to tackle environmental problems, and to be adaptable and responsive to new environmental demands and requirements. Customer EII provides valuable new information and knowledge that integrate with existing environmental management practices, allowing firms to manage the fluctuations in environmental landscape.

Lai et al. [8] showed that customer EII has a positive and significant effect on firm's profit and cost performance. Evidently, information sharing with customer on environmental-friendly products and processes enables firms to conduct better management plans in addressing environmental concerns. Customer EII is found as a useful source for cost saving and profit increase.

In the supply chain management, the importance of customer EII also suggests that environmental management capabilities originate from market intelligence collected through information sharing and communication with customers [40]. The study reveals that customer EII plays the strongest role among the three types of EII. The importance of customer EII can also be explained by rising market expectations and demands on environmental protection.

3.1.4.2 Internal EII

Wong [37] also finds that internal EII significantly contributes to corporate environmental adaptability, but not corporate environmental innovativeness. These findings imply that for firms to respond quickly to new environmental requirements internal EII is critical, but it has less impact on the ability to innovate new environmental products and practices. Internal EII provides an information sharing infrastructure, which improves cooperation across functions and the capabilities of firms to adapt to changes in organizational processes. This also points out the importance of having information acquisition mechanisms to enable knowledge creation and innovation processes across internal functions [41].

3.1.4.3 Supplier EII

As Wong [37] reveals, supplier EII does not contribute to corporate environmental innovativeness and adaptability. Although supplier EII enables efficient coordination with suppliers, it does not support innovation and responsive adaptations on environmental processes. The study finds that suppliers are important in providing information related to sustainable sources of inputs; however, they rarely play a strategic role in new product and process development for environmental protection of a focal firm. Without help from suppliers, firms need to design their new environmental management practices to ensure the new environmental management practices are compatible with their existing ones, and to avoid disruption to their operations.

Lai et al. [8] found that supplier EII can enhance the firm's performance related to cost and environment. In a highly munificent environment, supplier EII tends to improve firms' profits, cost, and environmental performance. Managers are called for to understand the importance of communicating with suppliers on the environmental standards and requirements. Since environmental munificence is found to play an important mediating role in determining the effects of EII on firm's performance, it is important to leverage supplier EII to improve cost-related and environmental-related performance under a high level of environmental munificence.

3.1.4.4 Cross-Influence of EII

Recent studies aim at identifying the relationship between internal EII and external EII. It is argued that that proactive manufactures will implement internal EII with an extension to their external partners such as suppliers and customers [42]. A focal organization tends to start improving its environmental performance by enhancing its internal EII, which requires the coordination and cooperation of different stakeholders within the organization. However, supply chain management is much more complex process which requires the coordination among the various outside participants including suppliers and customers. To comprehensively and sustainably improve the environmental performance, a focal organization has to increase its attention to the environmental performance of its suppliers and customers. Moreover, external EII is usually based on the successful commitment and support of internal EII implemented by the firm's senior managers [42]. The internal EII will subsequently facilitate firm's extension to adopt external EII. Supporting empirical evidences about this facilitation have been found in Japan as well as the automobile industry in Spain [43, 44].

3.1.5 Summary

There are three dimensions of EII, namely internal EII, supplier EII and customer EII. Internal EII supports the management of internal processes to achieve corporate environmental goals. It reflects how a firm organizes its communication and information sharing infrastructure across intra-organizational functions to maximize organizational efforts in environmental management. Managers should bear in mind that internal EII is important for firms in improving corporate environmental adaptability to the changing environmental demands and requirements. Firms can proactively develop internal and customer EII to increase organizational flexibility, in order to respond quickly to new environmental standards and customer expectations.

Recent research finds that supplier EII is helpful in enhancing firm's cost-related and environmental-related performance although it seldom affects the profit-related performance. This enhancement tends to more salient under a high-level of environmental munificence. One the other hand, customer EII tends to positively affect a firm's cost and profit-related performance, yet the effect has little difference in various levels of environmental munificence.

Supplier EII is concerned with the information sharing infrastructure that enables information sharing between a focal firm and its suppliers. Research finds that supplier EII does not contribute to corporate environmental innovativeness and adaptability. Customer EII, on the other hand, is concerned with the information sharing infrastructure of firms sharing environmental information with customers, while acquiring information of new customer requirements and expectations. It also facilitates business and environmental management opportunities identification and engaging customers in environmental protection activities. Research finds that customer EII is critical for environmental management capabilities. Managers should focus on strengthening their firms' customer EII to achieve a seamless information flow to develop environmental management capabilities. In case of limited resources, it is important to allocate resources to develop customer EII that provides an information sharing mechanism to facilitate firms in acquiring the latest market information for product or process improvement.

Internal EII and external EII are not two independent practices for a firm; rather, they are highly likely to be related. With the increasing requirements of environmentally-oriented management practices, a firm tends to improve the external EII upon the bases of the well-developed internal EII. Implementing internal EII will enhances the extent to which a firm adopts external EII.

3.2 Operational Adaptation

Adaptation of firms is the process of changing to become better suited for the business situation. In the supply chain context, adaption is referred to changing structure, strategy, or organization of a focal firm to meet new market needs and challenges together with its supply chain partners. The role of operational adaptation as a relational mechanism, with which firms are willing to adapt to the unforeseen needs of their partners in the supply chain, is an important aspect of inter-organizational coordination [45].

3.2.1 Inter-organizational Coordination

To facilitate coordination between firms in a supply chain, the following facilitating conditions, namely necessity, asymmetry, reciprocity, efficiency, stability and legitimacy [46], determine the formation of the inter-organizational relationships:

- *Necessity* is when organizations need to establish linkages or exchanges with supply chain partners to meet legal or regulatory requirements.
- *Asymmetry* refers to the potential to exercise power or control over another partner firm or its resources.
- *Reciprocity* refers to cooperation, collaboration, and coordination among supply chain partners focusing on pursuing common interest and goals.
- *Efficiency* refers to an organization's efforts to improve its internal productivity to fulfill its responsible supply chain activities.
- *Stability* refers to the adaptive response of organizations when facing environmental uncertainty in a supply chain.
- *Legitimacy* is devised from an organization attempt to demonstrate or improve its reputation, image, prestige, or agreement with prevailing norms in its supply chains.

Some organizations may minimize establishment of inter-organizational relationship with supply chain partners because it may result in losing some degree of freedom to act independently. It may also require investing scarce resources and energy to develop and maintain a relationship with another organization with unclear or intangible potential investment returns [47]. In the contrary, an organization may involve in an inter-organizational relationship when it perceives dependence on partners for accessing resources or specialized products or services [48]. Domain similarity determines whether organizations would engage in an inter-organizational relationship. Domain similarity defines the degree of which the organizations have the same services, clients, and personnel skills. Organizations that have moderately similar domains are likely to have complementary resources, thus they are likely to communicate frequently to achieve a mutually beneficial agreements [45]. However, domain similarities that are too high would hinder an inter-organizational relationship due to competition for similar resources.

3.2.2 Supplier Operational Adaptation

Supplier operational adaptation is the willingness of suppliers to meet their customer's requirements. It is a valuable relational mechanism for maintaining a stable and cooperative exchange relationship for economic gains [49]. In supply chain management and environmental management, supplier operational adaptation is regarded as the obligations and expectations in environmental protection between partner firms that occur through social processes in economic exchange [50]. It can also be viewed as a relational mechanism between partner firms that reinforces bilateral relationships and facilitates economic exchanges and environmental management practices to induce long-term orientation in buyer-supplier relationships [51]. Supplier operational adaptation is found to be crucial for firms to organize their production and marketing activities for quick response to market changes [52].

3.2.2.1 Drivers for Supplier Operational Adaptation

The General Theory of Network Governance is derived from the transaction cost economics and social network theory. It takes into account of both transactional and relational characteristics in economic exchange for coordinating business activities [53]. Transactional mechanism in a partner relationship concerns the nature of governance that emphasizes the costs of conducting exchange between firms (e.g., task coordination) [54], whereas relational mechanism explains the social interactions that could foster long-term exchange and trust between partner firms [55]. Business activities and environmental management practices would involve a specific transactional mechanism (e.g., information integration and exchange) that nurtures the relational mechanism (e.g., supplier operational adaptation) in a network of transacting partners, resulting in coordination efficiency [12]. This theory advocates that partner firms need to develop both transactional and relational mechanisms to govern economic exchanges to be prepared for unforeseen contingencies in their operating environments [53].

Enhancing a firm's operational flexibility through outsourcing and collaborating with partners can be useful to lower such business uncertainty as unstable customer demand and material supply [56]. The willingness of partner firms to handle the unforeseen changes, such as change of environmental regulations, customer preferences, and product specification, is required to foster operational flexibility in supply chain to cope with the changes [32]. Based on prior research, suppliers with more long-term customer relationships were found to outperform those with fewer; close relationships with customers allow suppliers to reduce inventory holding costs, and to reduce such discretionary expenses as administrative, selling, and overhead [57]. It has been found that the opportunity to learn from an innovative customer and supplier technical capability are strong drivers for suppliers to commit to collaborate with customers for environmental product development [58].

Product development and innovation drives collaboration between suppliers and buyers. For example, Boeing, an American manufacturer of aircraft, worked together with Nordam, an American aircraft component manufacturer, on developing a new composite window frame. The new window frame for the newly developed aircraft, 787 Dreamliner, weighs 50 % less than a traditional aluminum window frame, which is important to increase fuel efficiency and help achieving Boeing's environmental performance targets [59]. Moreover, investing in assisting a new environmental product development of a customer, who is a technology leader, enables a supplier to offer the newly developed know-how and technology to other customers. This offers potential to green the supply chain.

3.2.2.2 Forms of Supplier Operational Adaptation

Adaptation of environmental and supply chain management can be grouped into four forms: (1) product and process information exchange, (2) operational linkages to facilitate the flow of goods, services or information, (3) co-operative norms, and (4) relationship specific adaptations that are non-transferable [60]. Streamlined communication flow is of particular importance for supplier operational adaptation. Prior research suggests that at the beginning of buyer-supplier collaboration for environmental management, lack of understanding between partner firms might cause problems due to miscommunication. It is also crucial to maintain a close interaction and education relationship with the end-user [61], in order to understand their environmental expectations in new products and services. Firms that can leverage their supply chain collaboration and be tightly aligned in environmental management practices can become more innovative and efficient in achieving environmental protection goals. For example, Nestlé, a multinational food and beverage company, launched the Nestlé Cocoa Plan in 2009 to ensure a sustainable and high quality supply of cocoa for Nestlé in the long-term by working closely with farming cooperatives in Cote d'Ivoire, Ecuador and Venezuela [62]. Nestlé plans to invest CHF 110 million on cocoa creating shared value initiatives in 10 years. The plan concerns simplifying the supply chain, from farmers, cooperatives, manufacturers to consumers, so that returns to farmers would be improved as well as the quality of cocoa. Working with farmers, for example, involves training them in Ecuador and Côte d'Ivoire to help with increased yields, reduced disease, environmental sustainability and improved crop quality. Additionally, Nestlé has co-founded The World Cocoa Foundation (WCF) to tackle problems such as ineffective farming techniques and poor environmental management. By the end of 2010, some 340,000 children have benefited from WCF-supported education programmes and 8800 teachers have been trained.

By capitalizing on supplier operational adaptation, firms can be aware of the supplier inputs necessary for environmental management practices implementation in their supply chain to proceed. For example, Procter & Gamble Co. (P&G), an American multinational consumer goods company, launched the Supply Chain Environmental Sustainability Scorecard in 2010 to "improve the environmental footprint of P&G's supply chain, fuel innovation, and encourage suppliers to make environmental improvements in their own supply chains" [63]. One of the components, the Scorecard Analysis Tool was made publically available for any company to use in 2012. The Scorecard has enabled P&G to achieve improvements, increased innovation and collaboration within its supplier network. Some of the improvements include facilitating P&G and its suppliers to eliminate plastic windows on a brand's cartons, reducing manufacturing scrap waste, replacing petroleum-based materials with certified Roundtable on Sustainable Palm Oil (RSPO) material, and identifying opportunities to more efficiently transport products and reduce the quantity of automobiles used [63].

3.2.2.3 Impact of Supplier Operational Adaptation

When suppliers are willing to meet the customers' requirements and cooperate towards sustainable goals, operational adaptation can have a significant impact. Prior research on the manufacturing industry finds that sharing environmental knowledge with suppliers has a significant and positive effect on environmental performance. This is a form of investment in the supplier and to build relational capital. Relational capital is the total of actual and potential resources, which is inherent, available, or resulted from networks of relationships between organizations [64]. Value is created across an inter-organizational network and is synergized by the information, influence, and solidarity in the relationship between all parties [65]. A recent study finds that buyer's investment in the supplier builds relational capital, which brings forth mutual benefits. Relational capital has important mediating and moderating effects on the relationship between investment in the supplier and the benefits towards a focal firm [66].

The greening of suppliers would lead to improved environmental performance, competitiveness and economic performance [67]. Research on Chinese manufacturing enterprises finds that greening the supply chains, both internally and externally, involving collaboration and monitoring of suppliers can improve both environmental and economic performances. Additionally, quality management positively moderates the process [68]. This implicates that firms can focus on quality management to better improve the results from environmental management. For example, Unilever has successfully achieved better quality tea sourcing through sustainable practices. The company took steps to ensure product quality while improving sustainability at the same time with the help of suppliers. In 2007, Unilever started a large-scale project to sustainably source tea. Working with suppliers and heavily investing in improving farming practices among tea farmers, Unilever has managed to build capacity to more than 170,000 tonnes of Rainforest Alliance certified tea in five years [69].

In 2011, Unilever launched the 'Partner to Win' programme which focuses on building relationships with selected key supplier partners in order to achieve mutual growth. Under the programme, Unilever sets out 'Joint Business Development Plans' with specific partners, each plan identifies "Unilever's strategic business plans and provides a clear framework of how the two organisations will work together to deliver them over the long term." [70]. For example, Unilever has linked up with DHL, the multinational logistics company, in 2012 in a joint business development plan. This was a collaboration to reduce greenhouse gas emissions and to improve the environmental performance of both companies, with a "focus on cutting emissions, enhancing efficiencies across the supply chain, and reducing waste levels through the use of new technologies." [71]. The plan entailed that DHL would manage Unilever's logistics operations in emerging markets, which included plans to develop a blueprint for sustainable warehouse designs, methodologies for accurately measuring supply chain emissions, and strategies for reducing waste levels and boosting recycling rates [72]. Both companies would adopt a 'design once, deploy everywhere' strategy to accommodate the scale and breadth of the partnership and enable quick implementation of effective solutions [73]. That is, if a solution were proven effective being implemented in one country, it would be applied in the rest of the other countries as well. In 2013, Unilever has managed to reduce the energy used for manufacturing. It's CO_2 emissions from energy was 32 % below 2008 levels measured per tonne of production [74].

3.2.3 Summary

Supplier operational adaptation means the readiness of suppliers to meet their customer's requirements in order to maintain a stable and cooperative exchange relationship for economic gains. Suppliers with more long-term customer relationships tend to achieve better performance than those with fewer; close relationships with customers allow suppliers to reduce costs and expenses on inventory and overhead. The need to product development and innovation also drives collaboration between suppliers and buyers. The opportunity to learn from an innovative company is a strong driver for firms to commit to collaborating in a partner's product development.

There are several drivers for supplier operational adaptation. Partner firms need to develop transactional and relational mechanisms to regulate economic exchanges to be ready for contingencies in their operating environments. To improve a firm's operational flexibility, recent research finds that having a long-term supplier relationship would be beneficial, there would be a higher willingness for partner firms to accommodate the unforeseeable changes, such as new environmental regulations, customer demands, and product specification. Additionally, mutually investing in development and innovation also enables both parties to achieve a greener supply chain more efficiently.

Four forms of supplier operational adaptation include (1) product and process information exchange, (2) operational linkages to facilitate the flow of goods, services or information, (3) co-operative norms, and (4) relationship specific adaptations that are non-transferable. Efficient communication flow is vital, as a lack of understanding between partner firms would cause disruption in the flexibility and innovativeness of the collaboration.

From a focal firm's perspective, building relational capital is essential to gain environmental and economical improvements through supplier operational adaptation. Relational capital is the total value and resources that can be derived from networks of relationships between organizations. Recent research finds that through supplier operational adaptation and greening the supply chains can improve both environmental and economic performances. Quality management positively moderates this process, so managers should focus on quality management. In turn, improving environmental and economic performances as well as competitiveness.

3.3 Supplier Environmental Integration and Capability

There is increasing evidence suggesting that supply chain integration (SCI) has a positive impact on operational performance outcomes, such as for example lead-time, product quality, operational flexibility, and cost reduction [75]. SCI can be divided into three dimensions: internal, supplier, and customer integration [11]. The European Commission defines environmental integration as "making sure that environmental concerns are fully considered in the decisions and activities of other sectors." [76]. In the supply chain context, environmental integration means involving and taking account of environmental impact in such activities as procurement, production and delivery. Supplier environmental integration is concerned with strategic joint collaboration between a focal firm and its suppliers in managing the environmental issues in such cross-firm business processes as information sharing, strategic partnership, collaboration in planning, joint product development, and so forth [77].

Environmental management capability of suppliers is related to suppliers' ability to perform business activities in an environmentally friendly manner while attaining financial gains [78]. In general, it is seen as their ability to respond to the environmental concerns of the operational and environmental requirements of customers (i.e. buying firms) [79]. This capability is reflected by their adoption of an environmental management system standard (e.g. ISO 14000), environmental performance of their upstream suppliers, and environmental policy development to mitigate negative environmental impacts [80]. The environmental management capability of suppliers is important to the implementation of environmental management practices of a focal firm because 87 % of customers would accuse firms of environmental negligence when their suppliers are environmentally irresponsible, such as found using harmful chemicals, and refusal of product recycling [81]. If suppliers are convicted of using polluting chemicals, carcinogenic substances, and carbon emission processes, their downstream partners' (e.g., retailers) reputation can be damaged. Moreover, financial loss would be incurred, due to product recalls, legal fees, claims handling, and so forth [82]. First steps for focal firms to avoid the above losses are to factory audit and source from ISO 14000 certified suppliers to ensure supplier quality and their environmental capabilities [83].

3.3.1 Supplier Environmental Integration and Capability Example

3.3.1.1 Kimberly-Clark

Kimberly-Clark, an American personal care corporation, collaborates with its suppliers to work towards environmental goals. Acknowledging that suppliers play an important role towards the effectiveness of its environmental efforts, Kimberly-Clark developed the Supplier Social Compliance Standards Audit Guide in 2010. The guide outlines its sustainability expectations from suppliers, which includes environmental requirements and restrictions.

In addition to the guide, Kimberly-Clark started to work with its suppliers in 2010 to achieve one of its environmental sustainability goals, which was to use 100 % certified fiber [84]. Its procurement team engaged with fiber suppliers, and this goal was achieved in 2012 with its fiber suppliers providing 100 % third-party certified fiber [85]. Kimberly-Clark has also been working with its suppliers on reducing the environmental impact of packaging. In 2011, it started to collect data from its packaging suppliers worldwide to support the development of Kimberly-Clark packaging metric. In 2012, technical meetings were held with several of its major packaging suppliers to brainstorm ideas for sustainable packaging. In the same year, Kimberley-Clark implemented several packaging reduction projects across all businesses and regions. One of the results of these projects reduces the weight of its Kleenex facial tissue cartons, saving Kimerley-Clark nearly US$1 million [85].

3.3.1.2 Unilever

Unilever, an Anglo-Dutch multinational consumer goods company, is also committed to working with and supporting their suppliers of being environmentally sustainable. The Unilever Sustainable Agriculture Code was implemented in 2010, the code provides a guideline for its sustainable sourcing program through self-assessment or external certification standards. Working together with agricultural raw material suppliers, Unilever is striving to source 100 % of its agricultural raw materials sustainably by 2020. For example, the company has established the Knorr Sustainability Partnership Fund to support vegetable suppliers on sustainable agriculture projects [86].

In 2011, Unilever launched the Partner to Win programme to partner with 150 strategic partners for environmentally sustainable innovation. Through this program, Unilever shares its strategies and growth plans with suppliers. Since 2007, Unilever co-established a global Programme for Responsible Sourcing (AIM-PROGRESS). AIM-PROGRESS encourages consumer goods manufacturers like Unilever and suppliers to share, exchange, and promote responsible sourcing practices and sustainable production systems. Through this program, Unilever reaches out to suppliers to raise awareness of environmental issues. For example, through holding seminars with suppliers in Russia, India, South Africa, Mexico and Kenya, a number of key suppliers of Unilever re now adopting more responsible sourcing principles [87].

3.3.2 Impacts of Supplier Environmental Integration and Capability

According to prior studies, a low level of supplier and customer integration is likely to lead to inaccurate or distorted supply and demand information for a focal form, subsequently leading to poor production plans, high level of inventory, and poor delivery reliability [88]. External integration with supply chain partners improves process flexibility [89], as it enables supply chain partners to be able to better foresee and coordinate supply and demand [11]. This coordination flexibility is important for the improvement of delivery performance and responding to changing market demands.

When suppliers have the ability to be environmentally responsible in their environmental management practices such as product stewardship and process stewardship, it would have significant positive influence on pollution reduction by being green in the source of materials and product components [90]. The literature suggests that supplier environmental management capability is vital for developing environmental technologies [78], evaluating upper-tier supplier environmental performance [91], minimizing waste in business processes [92], and partnering with environmentally responsible upper-tier suppliers and service providers [93]. Working with suppliers with environmental management capability can build an environmentally friendly image and reduce cost relating to inspections, factory audits, and materials testing, especially for enterprises operating under a highly competitive and regulated industry [94]. Suppliers' environmental management capability also helps promoting environmental awareness and standards across the supply chain, from procurement through to transport and delivery [95].

In terms of asset recovery, suppliers play a role as well. Asset recovery concerns the separation and inspection of returned products and components for remanu-facturing and reuse, with a view to maximize the value of returned products. Suppliers contributes to environmental management practices such as recycling materials, remanufacturing, and redistributing products, where implementation is guided by firms' environmental management policies and systems [96]. Asset recovery involves a series of processes that cannot be fully planned in advance. Unforeseen situations arise due to different conditions and uncertainty of quantities of the returned product, and the capacity of the recovery processes [97]. Suppliers play a significant role in coping with these uncertainties by offering assistance in various environmental management practices.

Suppliers can affect the results of asset recovery depending on their involvement and participation [98]. Suppliers can participate in the recovery processes at the product component level, cooperating with manufacturers to refurbish components while safeguarding the recovered components' quality. Focal firms can lower their investment in asset recovery by outsourcing component recovery to suppliers, who have the expertise and capability to recondition components. This can also improve component design and production processes, as suppliers can help identify design flaws through participating in the component recovery process.

3.3.3 Summary

Supplier environmental integration concerns the strategic joint collaboration between a focal firm and its suppliers when managing environmental management practices such as product development, supply chain emission reduction and product recovery. Environmental management capability of suppliers is concerned with their ability to perform their business activities in an environmentally friendly manner while attaining financial gains. It can be represented by the adoption of an environmental management system standard (e.g. ISO 14000), and corporate environmental policy development. Supplier environmental integration also means that suppliers can participate in the asset recovery processes to cooperate with a focal firm to refurbish components and ensuring recovered components' quality.

Supplier integration improves process flexibility for firms, allowing firms to better predict and coordinate supply of components. This coordination flexibility is important for the improvement of delivery, lead-time, and responding to changing market demands in environmental management. Supplier environmental capability is also reflected in the products, production processes, business routines, sourcing, and communication, which all support environmental management practices of focal firms. Managers should source from suppliers who are ISO 14000 certified, conduct environmental evaluation on their second-tier suppliers, provide ecological proof of suppliers outputs, and communicate about their environmental manage-ment with supply chain partners.

3.4 Customer Environmental Integration

Customer environmental integration is related to the participation, attention and efforts of customers made to facilitate a firm's environmental practices. In supply chain integration, customer integration is concerned with the involvement of sharing strategic information and collaboration between a focal firm and its customers, thus with an aim to improve visibility and facilitates joint planning [99]. Customer environmental integration allows firms to have a deeper understanding of market expectations and opportunities, in turn offers products/services that satisfy customer environmental expectations and requirements [100].

Customer environmental information integration is referred to the organizational information sharing infrastructure for sharing environmental information with customers (e.g., new eco-product introduction), while collecting information about new market requirements and expectations (e.g., new environmental standards and regulations). This helps identify business and environmental opportunities [15], as well as informing, educating, and engaging customers to participating in various environmental protection activities, such as returning end-of-life products and choosing environmentally friendly alternatives.

Customer environmental integration is an integral part of extended producer responsibility (EPR) for firms to meet performance objectives. EPR is a policy concept aimed at extending producers' responsibility for their products to the post-consumption stage of product life with the prerequisite that the producers have the ability to reduce environmental impacts [101]. Under EPR, customer environmental integration facilitates participation of customers in the product return process that close the loops of product life-cycle and their attention and efforts made to facilitate the firm's EPR practices.

3.4.1 Channels of Customer Environmental Integration

3.4.1.1 Customer Collaboration

With the growth of Web 2.0 and social networks, customers' role are extending beyond solely purchasing and consuming services. Customers can also be involved in design, production, and marketing of products [102]. For example, Starbucks, a global coffee chain, launched MyStarbucksIdea.com in 2008. This is a social initiative that allows customers and partners to share their ideas with Starbucks. Users are also able to read other's ideas, discuss, and vote on ideas [103]. Submissions made via MyStarbucksIdea.com range from suggesting store openings, new flavors, product innovation, to environmentally friendly ideas. Examples of environmentally sustainable ideas include:

– Tempered glass bottle and protective silicone jacket [104]
– Reusable cup sleeve [105]

- Recycling through working with local government: In Seattle, for example, Starbucks is expected to divert 6000 tons of packaging (e.g., cups and lids) and food waste from a landfill to recycling each year. Starbucks has also increased the use of ceramic mugs for in-store drinking and customer tumblers. In 2009, nearly 1.2 million pounds of paper was this practice has achieved nearly 1.2 million pounds of paper from going into landfills [106].
- Food donation to addressing wasted pastries: Starbucks has collaborated with Food Donation Connection in 2010. This allows stores to donate unsold baked goods, packaged food items, and coffee to local organizations, helping to feed the community in need [107].

To achieve eco-efficiency and eco-effectiveness from product or service development, firms can start with having environmental issues on the agenda before developing market investigations, so that a customer focus would be involved in the beginning [108]. Zipcar, a car sharing company founded in 2000, aims to identify congestion and pollution problems, and finding sustainable solutions [109]. Zipcar crafts the experience from the customers' point of view through developing an application that allows car booking and unlocking on the mobile anywhere any time [110]. Zipcar has already achieved several green benefits, including taking out 15 cars off the road for each Zipcar, and reducing 5500 miles drove per year for 90 % of the Zipcar members, which is equivalent to 32 million gallons of crude oil saved. Zipcar also continuously works with carmakers for environmentally friendly automobile alternatives such as plug-in vehicles. In 2011, Toyota delivered eight plug-in Toyota Prius hybrids to Zipcar, before it is publicly sold in the market in 2012 [111]. Zipcar has also worked with Vauxhall in 2012 and brought the first electric car to the market. This partnership was to strive towards the company's goal of making positive changes to the city environment [112].

3.4.1.2 Environmental Information Sharing

In some occasions, customers would associate sustainability with sacrifices of comfort, time, and money. Customers' acceptance of the trade-offs is a precondition to achieving sustainability [113]. In order to achieve this, sharing of environmental information with customers is crucial. Lush, a handmade cosmetics company in the United Kingdom is no stranger to environmental protection. On the company website and in store publications, the company informs customer about its efforts in such environmental policies as ethical sourcing and minimal product packaging In addition, Lush stopped using palm oil, which is a common ingredient used in soap. The high demand of palm oil is causing negative effects on the Indonesian rainforest and ecosystem. Thus, Lush launched the Jungle soap collection that does not use palm oil. To celebrate this new soap base and to educate customers about palm oil, Lunch has covered all shop windows in green palm prints and sold the Jungle soap to raise funds for Rainforest Action Network [114]. In 2007 and 2008, Lush launched "Naked" in-store campaign to raise awareness of the negative environmental impact of unnecessary packaging. Shop assistants were wearing nothing but an apron with the "Ask me why I'm naked" slogan print [115].

3.4.1.3 Customer Involvement in Product Returns

Post-consumer stage of the product life-cycle is of raising importance [116]. Asset recovery contributes to environmental management through practices such as recycling materials, remanufacturing, and redistributing products [96]. Lush has launched a take back program to encourage its customers to return Lush block pots to Lush's store to receive a free fresh facemask. Lush blog posts are constantly giving advices to customers about being green. It also shares and teaches customers how the black pots packaging can be reused and recycled into floor tiles in its website [117]. This allows the residual values of the black pots to be fully utilized by grinding up and melting old pots to make into new pots [118]. Similarly, Hewlett-Packard Company (HP), an American multinational information technology corporation, forms a "closed-loop" recycling process, with more than 1.5 billion HP LaserJet and ink cartridges produced containing content from old cartridges. Cartridges and parts can be reduced to raw materials to make new cartridges and other metal and plastic products [119]. HP also encourages customers to recycle, trade-in, return for cash, donate, and properly disposal used products by providing the information on its website.

3.4.1.4 Quality Function Deployment

Quality function deployment (QFD) is a tool used for translating the 'voice' of customers into technical requirements such that internal processes can deliver what the customers want and meet their requirements [120]. By this definition, QFD is a customer-driven technique for companies to design and plan their offerings of product and services. It is widely used in practical applications such as engineering, strategy development, and supplier evaluation and selection [121, 122]. The QFD process starts with constructing a "house-of-quality" which translates customer requirements into design requirements and subsequently deploy such requirements to various organizational functions to meet the needs of customers [120].

Prior research in management relates QFD to environmental concerns such as green product design and development [123]. With the growing awareness in low-carbon economic development, the adoption of green shipping practices (GSP) has been gaining increasing popularity in shipping industry. Recent study focuses on the use of QFD for customer EII in shipping operations [120]. To better fulfill the customer quests for environmental sustainability, shipping companies should understand the environmental expectation of their customers (shippers) and decide how to translate those expectations into their service processes [120]. Lam and Lai (2014) conducted a case study of an international tanker shipping company to provide insights on "what" and "how" to satisfy the environmental requirements of customers. They found that the use of green design vessels, engines, and machinery is the most important element that can respond most effectively to the customer requirements of mitigating carbon emissions and pollution and efficient use of fuels. In addition, the use of green design vessels is found desirable for other design requirements such as the use of eco-friendly paints and option to alternative energy resources. Their analysis suggests that shipping companies should give priority to the use of green design vessels.

3.4.2 Impacts of Customer Environmental Integration

Recent research finds that customer environmental integration influences the relationship between EPR practices and the market and financial performance outcomes [124]. Firms would need customer integration to bring the performance benefits of EPR. Customer integration provides essential market information on the latest environmental protection requirements and expectations, allowing firms to strategize to cater market needs. Thus, a higher level of customer integration can be beneficial for improving corporate image and organizational positioning. However, customer integration would introduce uncertainty to the implementation of EPR. The uncertainty was induced by the dependence of customer willingness to cooperate and facilitate for the amount and delivery time of returned products fluctuates [125]. Additionally, customers would scrutinize the performance outcomes of the EPR practices as customer integration increases. All in all, these uncertainties and expectations would introduce excessive or shortage of returned goods and incur costs to firms due to extra organizational efforts.

Recent research finds that customer environmental information integration is complimentary to the environmental management capabilities of firms, particularly through effective information sharing and communication with customers [37]. Customer environmental information integration provides information about market needs, in which reduces uncertainties in eco-product design, asset recovery planning, and reverse logistics scheduling. As a result, customer environmental information integration enhances organizational adaptability to market changes [89], and improving a firm's responsiveness towards unforeseeable market demands and changes in environmental regulations [11]. Therefore, the mutual exchange of information can facilitate a collaborative relationship between firms and customers on identifying environmentally beneficial market opportunities and meeting market needs.

Stakeholder support for asset recovery is essential as they can exert pressure that influences organizational environmental decisions and practices [126]. Customers, which belong to the stakeholder group that influence the frontline environmental management processes of firms, can affect the performance of asset recovery [98]. Research findings suggest that stakeholder support strengthens asset recovery's ability to improve product quality and financial performance [127]. Customers provide end-of-life, end-of-use, or unused products back to firms through reverse logistics chains are useful as the input for asset recovery practices.

3.4.3 Summary

Customer environmental integration is concerned with the involvement of customers, which include their participation, awareness, and attention, when a firm is implementing environmental practices. There are several channels of implementing costumer environmental integration, including customer collaboration, environmental information sharing, customer participation in product returns, and quality function deployment. The sharing of strategic information and collaboration between a firm and its customers, with an aim to improve visibility and facilitates joint planning, is the principle of customer integration in a supply chain. Customer environmental information integration facilitates firms to collect information about new market requirements and expectations, which is critical to environmental management capabilities of firms.

Customer's participation in the product return process facilitates the firm's EPR practices and asset recovery processes. Customer integration influences the relationship between EPR practices and the market and financial performance outcomes. When customer integration is high, the positive relationship between implementing EPR practices and market performance is strengthened, whereas a high level of customer integration would have detrimental effects on the relationship between EPR practices and financial performance. Thus, managers should control the cost of coordinating product returns from customers so that it does not compromise the financial performance gains from EPR practices.

3.5 Consumer Preferences

With green advertising of eco-products and social expectations today, purchase decision of customers often involve making an environmentally sustainable or unsustainable choice. Consumers are interested in being green, however personal choices and needs do add up. Thus, companies need to gain a comprehensive understanding of customer preferences and attitude towards sustainable consumption in order to meet customer needs.

3.5.1 Sustainable Consumption

The Oslo Roundtable on Sustainable Production and Consumption in 1994 defined *sustainable consumption* as "the use of goods and services that respond to basic needs and bring a better quality of life, while minimizing the use of natural resources, toxic materials and emissions of waste and pollutants over the life cycle, so as not to jeopardize the needs of future generations." [128]. To gain an understanding of sustainable consumption, National Geographic has partnered with GlobeScan to launch Greendex, which is a research to collect and monitor consumer progress toward environmentally sustainable consumption since 2008. In 2012, a total of 17,000 consumers in 17 countries were studied. The consumers were interviewed about such behaviors as energy use and conservation, transportation choices, food sources, the relative use of green products versus conventional products, attitudes toward the environment and sustainability, and knowledge of environmental issues [129]. The "Greendex 2012: Consumer Choice and the Environment—A Worldwide Tracking Survey" finds that environmentally friendly consumer behavior has increased from 2010 only in 5 out of 17 countries. This progress though positive, the momentum of sustainable consumption has a downward trend. In regards to goods consumption, the majorities of consumers in 16 out of 17 countries prefer to repair rather than replace [130]. Additionally, consumers in most countries prefer reusable to disposable products. However, price premium of green products remains an obstacle of sustainable consumption. Many consumers are undecided on sacrifices money for environmentally friendly products [130]. Yet, the survey shows that percentages willing to pay a premium have trended upward since 2008.

3.5.2 The Green Consumer

Green consumers show "an interest in product's characteristics such as recyclability and chemical content, favorably discriminating consumption towards products that are organic, energy efficient or have biodegradable packaging [131]." The International Institute for Sustainable Development has profiled the 'best' green consumers as young adults with relatively more disposable income, better education, and intellectual orientation [132]. Coincide with this profiling, according to a survey of 28,000 consumers in 56 countries conducted by Nielsen in 2012, 66 % of global consumers prefer to buy products and services from companies that give back to society, out of which 63 % are under the age of 40 [133]. Nielsen also defines these socially conscious consumers as those who are willing to pay a premium for sustainable products. The survey further identifies that these socially conscious consumers believe that environmental sustainability is a cause that companies should support.

Prior research defines three types of green consumers by separating the green consumption process into stages and considering their environmental orientation [134]:

- "Grey consumers" are those who are skeptical about environmentalists' claims and have little interest in the environment

– "Consistent ecologists" are constantly influenced and informed by environmental concerns, which reflects on their lifestyle, purchasing, and consumption.
– "Fit and forget" green consumers are those who would make green purchases without intentions to change their lifestyles or consumption and disposal patterns.

3.5.3 Green Consumer Preference and Behavior

The attitude-behavior gap remains a concern for eco-product markets. Consumers care about the environment, however there are difficulties to align their behaviors with their intentions. While majority of consumers are aware of the importance of sustainable consumption, they often overstate their green behavior and claim to be willing to make sacrifices. They also tend to be skeptical about companies' environmental claims. It is similar to how people often fail to act on intentions to diet or exercise, we know it is for the best but the slight sacrifices put off actions. For example, prior study reveals that fashion consumers are interested in purchasing eco-fashion but are not willing to sacrifice personally, such as paying a higher price [135]. According to Nielsen's 2011 Global Online Environment AND Sustainability Survey of more than 25,000 consumers in 51 countries, only 22 % of global online consumers are willing to pay more for eco-friendly products [136]. Nonetheless, there is an upward trend of change. The Nielsen Global Survey of Consumer Shopping Behavior 2013 finds that an average of 46 % of global consumers would purchase environmentally friendly products regardless of price [137].

However, when it comes to eco-fashion, green apparel still remains a niche retail market and consumer segment. The Cotton Incorporated Retail Monitor survey finds that environmentally friendly apparel are priced 7 % higher on average than those not marketed as green in 2010, and just over a third of consumers would be willing to pay the premium [138]. So, what is the tipping point and what triggers the consumers decide to purchase green apparel? Green consumer consumption encapsulates the following stages of the consumer decision-making process: information search, evaluation of alternatives, and purchase decisions. Prior research on eco-fashion consumption decision (ECD) defines store-related attributes (SRA) influence ECD the most among other factors [139].

Store related attributes (SRA) are the characteristics of the store, which are related to the store's features and store operations [140]. Fashion consumers would observe customer service, store design and environment, store's ethical practices, and shop convenience as common SRA [141]. The consumer decision-making process (CDMP) [142] theory advocates that a fashion consumer would evaluate various product related attributes (PRA) and SRA when recognizing the need for eco-fashion, before making a final purchase decision. A recent study finds that SRA positively influence consumers' eco-fashion consumption decision [143]. When fashion consumers perceive a higher level of customer service quality, they are more satisfied with the retail store and its offerings, thus leading to a higher chance of purchase [144]. A pleasant store environment also stimulates fashion consumer purchase decision [145]. Store atmosphere such as the ambient factor (e.g., lighting and music) and social factors (e.g., friendliness of sales associates) also have an influence on consumer purchase decisions [146].

Retail stores are also increasingly integrating ethical practices into their operations to be responsible to the environment [147]. For example offering recycling service and recyclable products can enhance fashion consumers' perceived effectiveness of environmental protection [148]. H&M, the Swedish multinational clothing retailer known for fast fashion, was the first fashion company to launch a global initiative called "Garment Collecting", to encourage customers to take unwanted garments of any brand and in any condition to H&M stores for rewearing, reusing, and recycling. As incentive, customers would be given a small amount of discount in return. For example, H&M offers a voucher of 15 % off one item per bag of old garments in the U.S. [149]. Fashion consumers would perceive that by purchasing at a store that implement environmental management practices could indirectly help protect the environment [150]. Thus, this program was deemed a success because not only it gives shoppers incentives to make more purchases, but also puts forth a sustainable image of H&M because the retailer has collected 7.7 million pounds of used clothing to good causes. H&M also donates USD 3 cents for every two pounds of collected clothing [151]. Therefore, SRA would have a positive influence on fashion consumers' purchase decision.

3.5.4 Impact of Price Premium

Price premium is referred to the pricing that is above the average price [152]. As aforementioned, the price premium of eco-products is a main put off for consumers' green purchases. According to an OgilvyEarth study, the price premium of sustainable products is the main reason that consumers do not make a purchase [153]. According to the conventional economic theory [154], price premium represents a monetary measure of what is sacrificed in their eco-product purchase.

Similarly to all consumers, fashion consumers would not be willing to pay a premium for eco-fashion as they often consider it is not value for money [152]. It is suggested that a 10 % price premium is within an acceptable range for fashion consumers and would not affect their willingness to purchase, but a 25–30 % price premium would have a negative effect [155]. The conventional economic theory [154] also suggests that when the price level of fashion clothing is high and the consumer is less familiar with the clothing, the consumer would perceive a risk of an incorrect assessment [156]. Thus, fashion consumers would use price as an indicator of quality in such circumstances, thus a greater willingness to buy. Therefore, the price level of fashion clothing moderates the perceived quality of the product, which in turn affects consumer willingness to make a purchase. Price premium of eco-fashion is also found to have a negative effect on the relationship between SRA and fashion consumers' purchase decision [143]. Customer service and stores' ethical practices are relatively intangible and difficult to assess [147], thus customers would use the price premium to evaluate SRA when making their purchase decision.

3.5.5 Summary

Sustainable consumption is purchasing goods or services that fulfill needs of a consumer, while minimizing the negative impacts on the environment, such as use of natural resources, toxic materials, and emissions of waste and pollution. Consumers in general care about the environment and are interested in sustainable consumption so that future generations can live a better life. Sustainable consumption behavior can include energy use and conservation, transportation choices, food sources, the relative use of green products versus conventional products, attitudes toward the environment and sustainability, and knowledge of environmental issues.

Green consumers are those who are interested in a product's sustainable characteristics such as its recyclability and chemical content. They would also prefer organic and energy efficient products with biodegradable packaging. The green consumer is profiled as young adults with relatively more disposable income, better education and intellectual orientation. However, there is an attitude-behavior gap with regards to green consumers. A majority of consumers are aware of the importance of sustainable consumption, they can be unwilling to pay premium price for sustainable products.

References

1. Madsen PM (2009) Does corporate investment drive a "race to the bottom" in environmental protection? A reexamination of the effect of environmental regulation on investment. Acad Manag J 52(6):1297–1318
2. Parmigiani A, Klassen RD, Russo MV (2011) Efficiency meets accountability: performance implications of supply chain configuration, control and capabilities. J Oper Manag 29 (3):212–223
3. ATKearney, *Cargon Disclosure Project Supply Chain Report 2011*, in *Carbon Disclosure Project*. 2011
4. Almotairi B, Lumsden K (2009) Port logistics playform integration in supply chain management. Int J Shipping Transp Logistics 1(2):194–210
5. Lai K-H, Wong CWY, Cheng TCE (2010) Bundling digitized logistics activities and its performance implications. Ind Mark Manage 39(2):273–286
6. Elliot S (2011) Transdisciplinary perspectives on environmental sustainability: a resource base and framework for IT-enabled business transformation. MIS Q 35(1):197–236
7. Erlandsson J, Tillman AM (2009) Analysing influencing factors of corporate environmental information collection, management and communication. J Clean Prod 17:800–810

8. Lai K-h, Wong CWY, Lam JSL (2014) Sharing environmental management information with supply chain partners and the performance contingencies on environmental munificence. Int J Prod Econ (In Press, Corrected Proof)
9. Green K, Morton B, Now S (1996) Purchasing and environmental management: interactions policies and opportunities. Bus Strategy Environ 5(2):188–197
10. Astley WG, Van de Ven AH (1983) Central perspectives and debates in organization theory. Adm Sci Q 28(2):245–273
11. Flynn BB, Huo B, Zhao X (2010) The impact of supply chain integration on performance: a contingency and configuration approach. J Oper Manag 28(1):58–71
12. Lai K-H, Bao Y, Li X (2008) Channel relationship and business uncertainty: evidence from the Hong Kong market. Ind Mark Manage 37(6):713–724
13. Inditex (2012) Annual report, p 322
14. Inditex (2014) Sustainable Inditex 2011–2015 (cited 4 Feb 2014). Available from: http://www.inditex.com/documents/10279/28230/Grupo_INDITEX_plan_1115_en.pdf/ec2e49f0-5cf5-4e7b-b597-b72fc8a0cb7a
15. Priem RL, Swink M (2012) A demand-side perspective on supply chain management. J Supply Chain Manag 48(2):7–13
16. Inditex (2014) Inditex's zero discharge pledge (cited 4 Feb 2014). Available from: http://www.inditex.com/en/sustainability/environment/manufacturing
17. Levine SR (2013) How Zara took customer focus to new heights. Credit Union Times
18. Dwyer J (2010) A clothing clearance where more than just the prices have been slashed (cited 19 Feb 2014). Available from: http://www.nytimes.com/2010/01/06/nyregion/06about.html?partner=rss&emc=rss&_r=1&
19. Inditex (2001) Code of conduct for manufacturers and suppliers (cited 6 Feb 2014]. Available from: https://www.inditex.com/documents/10279/28230/Grupo_INDITEX_codigo-de-co-nducta-de-fabricantes-y-proveedores_ENG.pdf/ade5106d-f46a-487b-a269-60c2e35cdcf4
20. Theyel G (2006) Customer and supplier relations for environmental performance. In: Sarkis J (ed) Greening the supply chain. Springer, London
21. Theyel G (2000) Management practices for environmental innovation and performance. Int J Oper Prod Manag 20(2):249–266
22. Gunther M (2012) Behind the scenes at the sustainable apparel coalition. In: GreenBiz.com
23. Coalition, S.A. Overview. 2012 [cited 2014 February 11]; Available from: http://www.apparelcoalition.org/overview/
24. Kibbey J (2014) Sustainable fashion: changing the rules of collective action. In: Guardian sustainable business 2014. The Guardian
25. Project CD (2014) Supply chain program (cited 11 Feb 2014). Available from: https://www.cdp.net/en-us/programmes/pages/cdp-supply-chain.aspx
26. Project CD (2014) Supply chain report 2013–14, p 35
27. Romano P (2003) Coordination and integration mechanisms to manage logistics processes across supply networks. J Purch Supply Manag 9(3):119–134
28. Wong CWY, Lai K-H, Ngai EWT (2009) The role of supplier operational adaptation on the performance of IT-enabled transport logistics under environmental uncertainty. Int J Prod Econ 122(1):47–55
29. Huo BF, Zhao XD, Zhou HG (2014) The effects of competitive environment on supply chain information sharing and performance: an empirical study in China. Prod Oper Manag 23:552–569
30. Subramani M (2004) How do suppliers benefit from information technology use in supply chain relationships? MIS Q 28(1):45–73
31. Stump RL, Sriram V (1997) Emplying information technology in purchasing: buyer-supplier relationships and size of the supplier base. Ind Mark Manage 26(2):127–136
32. Hallen L, Johanson J, Seyed-Mohamed N (1991) Interfirm adaptation in business relationships. J Marketing 55(1):29–37
33. Zenger TR, Hestlerly WS (1997) The disaggregation of U.S. corporations: selective intervention, high-power incentives, and molecular units. Organ Sci 8(2):209–222

34. Attaran M, Attaran S (2007) Collaborative supply chain management: the most promising practice for building efficient and sustainable supply chains. Bus Process Manag J 13(3): 390–404

35. Lee HL, So KC, Tang CS (2000) The value of information sharing in a two-level supply chain. Manage Sci 46:626–643

36. Lai K-h, Wong CWY, Venus Lun YH (2014) The role of customer integration in extended producer responsibility: a study of Chinese export manufacturers. Int J Prod Econ 147: 284–293

37. Wong CWY (2013) Leveraging environmental information integration to enable environmental management capabilities and performance. J Supply Chain Manag 49 (2):114–136

38. Zhao L et al (2013) The impact of supply chain risk on supply chain integration and company performance: a global investigation. Supply Chain Manag Int J 18(2):115–131

39. Geffen CA, Rothenberg S (2000) Suppliers and environmental innovation: the automotive paint process. Int J Oper Prod Manag 20(2):166–186

40. Tate WL, Ellram LM, Kirchoff JF (2010) Corporate social responsbility reports: a thematic analysis related to supply chain management. J Supply Chain Manag 46(1):19–37

41. Helfat CE (1997) Know-how and asset complementarity and dynamic capability accumulation. Strateg Manag J 18(5):339–360

42. Zhu Q, Sarkis J, Lai K (2013) Institutional-based antecedents and performance outcomes of internal and external green supply chain management practices. J Purch Supply Manag 19:106–117

43. Zhu Q, Geng Y, Fujita T, Hashimoto S (2010) Green supply chain management in leading manufacturers: case studies in Japanese large companies. Manag Res Rev 33(4):380–392

44. Gonzalez P, Sarkis J, Adenso-Diaz B (2008) Environmental management system certification and its influence on corporate practices evidence from the automotive industry. Int J Oper Prod Manag 28(11–12):1021–1041

45. Walter A (2003) Relationship-specific factors influencing supplier involvement in customer new product development. J Bus Res 56(9):721–733

46. Oliver C (1990) Determinants of interorganizational relationships: integration and future directions. Acad Manag Rev 15(2):241–265

47. Van de Ven AH, Walker G (1984) The dynamics of interorganizational coordination. Adm Sci Q 29(4):598–621

48. Cook KS (1977) Exchange and power in networks of interorganizational relations. Sociol Q 18(1):62–82

49. Cannon JP, Achrol RS, Gundlach GT (2000) Contracts, norms, and plural form governance. J Acad Mark Sci 28(2):180–194

50. Poppo L, Zenger T (2002) Do formal contracts and relational governance function as substitutes or complements. Strateg Manag J 23(8):707–725

51. Yang J et al (2008) Relational stability and alliance performance in supply chain. Omega 36 (4):600–608

52. Kelle P, Miller PA, Akbulut AY (2007) Coordinating ordering/shipment policy for buyer and supplier: numerical and empirical analysis of influencing factors. Int J Prod Econ 108 (1–2):100–110

53. Jones C, Hesterly WS, Borgatti SP (1997) A general theory of network governance: exchange conditions and social mechanisms. Acad Manag Rev 22(4):911–945

54. Williamson OE (1975) Markets and hierarchies: analysis and antitrust implications. The Free Press, New York

55. Uzzi B (1997) Social structure and competition in interfirm networks: the paradox of embeddedness. Adm Sci Q 42(1):35–67

56. Milliken FJ (1987) Three types of perceived uncertainty about the environment: state, effect, and response uncertainty. Acad Manag Rev 12(1):133–143

57. Kalwani MU, Narayandas N (1995) Long-term supplier relationships: do they pay off for supplier firms. J Mark 50(1):1–16

58. LaBahn DW, Krapfel R (2000) Early supplier involvement in customer new product development: a contingency model of component supplier intentions. J Bus Res 47(3): 173–190
59. Group N (2007) Nordam pioneers composite window frames for Boeing 787 Dreamliner (cited 18 Feb 2014). Available from: http://www.nordam.com/news/Lists/Press%20Releases/DispForm.aspx?ID=25
60. Davies T (2006) Collaborate to innovate (cited 14 Feb 2014). Available from: http://www.supplymanagement.com/resources/2006/collaborate-to-innovate
61. Rönnberg-Sjödin D (2013) A lifecycle perspective on buyer-supplier collaboration in process development projects. J Manufact Technol Manag 24(2):235–256
62. Nestlé (2012) The cocoa plan (cited 15 Apr 2014). Available from: http://www.nestle.com/csv/case-studies/allcasestudies/pages/the-cocoa-plan.aspx
63. Gamble P (2012) P&G shares data, results and analysis tool from supplier scorecard 2012 (cited 15 Apr 2014). Available from: http://news.pg.com/press-release/pg-corporate-announcements/pg-shares-data-results-and-analysis-tool-supplier-scorecard
64. Nahapiet JE, Ghoshal S (1998) Social capital, intellectual capital, and the organizational advantage. Acad Manag Rev 23(2):242–266
65. Adler PS, Kwon S (2002) Social capital: prospects for a new concept. Acad Manag Rev 27 (1):17–40
66. Blonska A et al (2013) Decomposing the effect of supplier development on relationship benefits: the role of relational capital. Ind Mark Manage 42(8):1295–1306
67. Rao P (2002) Greening the supply chain: a new initiative in South East Asia. Int J Oper Prod Manag 22(6):632–655
68. Sarkis J, Zhu QH (2004) Relationships between operational practices and performance among early adopters of green supply chain management practices in Chinese manufacturing enterprises. J Oper Manag 22:265–289
69. Unilever (2014) Sustainable living targets & performance (cited 24 Apr 2014). Available from: http://www.unilever.com/sustainable-living/sustainablesourcing/targets/
70. Unilever (2012) Long term supplier partnerships key to Unilever sustainable growth strategy (cited 23 Apr 2014). Available from: http://www.unilever.com/mediacentre/pressreleases/2012/longtermsupplierpartnershipskey.aspx
71. Murray J (2012) Unilever, DHL extend carbon-cutting partnership (cited 24 Apr 2014). Available from: http://www.greenbiz.com/news/2012/08/10/unilever-dhl-carbon-partnership
72. King B (2012) Unilever, DHL expand relationship with focus on supply chain efficiency (cited 24 Apr 2014). Available from: http://www.sustainablebrands.com/news_and_views/articles/unilever-dhl-expand-relationship-focus-supply-chain-efficiency
73. Leach A (2012) Unilever links up with DHL in joint business development plan (cited 24 Apr 2014). Available from: http://www.supplymanagement.com/news/2012/unilever-links-up-with-dhl-in-joint-business-development-plan
74. Unilever (2013) Unilever annual report and accounts 2013, p 147
75. Souder WE, Sherman JD, Davies-Cooper R (1998) Environmental uncertainty, organizational integration, and new product development effectiveness: a test of contingency theory. J Prod Innov Manage 15(6):520–533
76. Commission E (2014) Environmental integration (cited 19 Feb 2014). Available from: http://ec.europa.eu/environment/integration/integration.htm
77. Ragatz GL, Handfield RB, Petersen KJ (2002) Benefits associated with supplier integration into new product development under conditions of technology uncertainty. J Bus Res 55 (3):389–400
78. Klassen RD, Vachon S (2003) Collaboration and evaluation in the supply chain: the impact on plant-level environmental investment. Prod Oper Manag 12(3):336–352
79. Lu LYY, Wu CH, Kuo TC (2007) Environmental principles applicable to green supplier evaluation by using multi-objective decision analysis. Int J Prod Res 45(18–19):4317–4331
80. Corbett CJ, Kirsch DA (2001) International diffusion of ISO 14000 certification. Prod Oper Manag 10(3):327–342

81. Cotton Incorporated (2008) Lifestyle Monitor Trend Magazines
82. Economy E, Liebertha K (2007) Scorched earth: will environmental risks in China overwhelm its opportunities? Harv Bus Rev 85(6):88–96
83. King AA, Lenox MJ, Terlaak A (2005) Strategic use of decentralized institutions: exploring certification with the ISO14001 management standard. Acad Manag J 48(6):1091–1106
84. Kimberly-Clark (2010) Sustainability report 2010: collaborating with our suppliers (cited 19 Feb 2014). Available from: http://www.sustainabilityreport2010.kimberly-clark.com/people/collaborating-with-our-suppliers.asp
85. Kimberly-Clark (2012) Sustainability report 2012 (cited 19 Feb 2014); Available from: http://www.sustainabilityreport2012.kimberly-clark.com/our-approach/about-this-report/index.aspx
86. Unilever (2014) Working with suppliers (cited 19 Feb 2014); Available from: http://www.unilever.com/sustainable-living-2014/reducing-environmental-impact/sustainable-sourcing/sustainable-palm-oil/working-with-suppliers/
87. Unilever (2014) Advancing Human Rights with Suppliers (cited 19 Feb 2014). Available from: http://www.unilever.com/sustainable-living-2014/enhancing-livelihoods/fairness-in-the-workplace/advancing-human-rights-with-suppliers/
88. Lee HL, Padmanabhan V, Whang S (1997) The bullwhip effect in supply chains. Sloan Manag Rev 38(3):93–102
89. Ettle JE, Reza E (1992) Organizational integration and process innovation. Acad Manag J 34:795–827
90. Wong CWY et al (2012) Green operations and the moderating role of environmental management capability of suppliers on manufacturing firm performance. Int J Prod Econ 140 (1):283–294
91. Min H, Galle WP (2001) Green purchasing purchases of US firms. Int J Oper Prod Manag 21 (9):1222–1238
92. Michaelis P (1995) Product stewardship, waste minimization and economic efficiency: lessons from Germany. J Env Plann Manag 38(2):231–243
93. Seuring S (2004) Integrated chain management and supply chain management comparative analysis and illustrative cases. J Clean Prod 12(4):1059–1071
94. Tan JJ, Litschert RJ (1994) Environment-strategy relationship and its performance implications: an empirical study of the Chinese electronics industry. Strateg Manag J 15 (1):1994
95. Russo MV, Fouts PA (1997) A resource-based perspective on corporate environmental performance and profitability. Acad Manag J 40(3):534–559
96. Melnyk SA, Sroufe RP, Calantone R (2003) Assessing the impact of environmental management systems on corporate and environmental performance. J Oper Manag 21 (3):329–351
97. Corbett CJ, Klassen RD (2006) Extending the horizons: environmental excellence as key to improve operations. Manuf Serv Oper Manag 8(1):5–22
98. Polonsky MJ, Ottman J (1998) Stakeholders' contribution to the green new product development process. J Mark Manag 14:533–557
99. Fisher ML et al (1994) Making supply meet demand in an uncertain world. Harvard Bus Rev 72(3):83–93
100. Swink M, Narasimhan R, Wang C (2007) Managing beyond the factory walls: effects of four types of strategic integration on manufacturing plant performance. J Oper Manag 25: 148–164
101. OECD (2001) Extended producer responsibility: a guidance manual for governments. OECD Editor, Paris
102. Sigala M (2009) E-service quality and Web 2.0: expanding quality models to include customer participation and inter-customer support. Serv Ind J 29(10):1341–1358
103. Corporation S (2013) My Starbucks idea (cited 4 Mar 2014). Available from: http://mystarbucksidea.force.com/

104. Staff S (2011) New year, new merchandise (cited 4 Mar 2014). Available from: http://blogs.starbucks.com/blogs/customer/archive/2011/01/14/new-year-new-merchandise.aspx
105. Staff S (2010) Reusable Cup Sleeve: like a fall sweater for your favorite warm beverage! (cited 4 Mar 2014). Available from: http://blogs.starbucks.com/blogs/customer/archive/2010/09/07/reusable-cup-sleeve-like-a-fall-sweater-for-your-favorite-warm-beverage.aspx
106. Staff S (2010) Progress report on a top idea: recycling (cited 4 Mar 2014). Available from: http://blogs.starbucks.com/blogs/customer/archive/2010/07/06/progress-report-on-top-idea-recycling.aspx
107. Staff S (2010) Idea launched: stop wasting so many pastries (cited 4 Mar 2014). Available from: http://blogs.starbucks.com/blogs/customer/archive/2010/03/15/idea-launched-stop-wasting-so-many-pastries.aspx
108. Ritzén S (2000) Integrating environmental aspect into product development—proactive measures. Department of Machine Design Integrated Product Development division, Royal Institute of Technology, Stockholm, Sweden
109. Zipcar I (2014) About us (cited 2014 16 Mar 2014). Available from: http://www.zipcar.com/about
110. Limited E (2014) Zipcar shapes up with design to deliver a delightful customer experience (cited 29 Apr 2014). Available from: http://www.eyefortravel.com/social-media-and-marketing/zipcar-shapes-design-deliver-delightful-customer-experience
111. Schultz J (2011) Zipcar adds Toyota Prius plug-in hybrids to fleet (cited 4 Mar 2014). Available from: http://wheels.blogs.nytimes.com/2011/01/27/zipcar-adds-toyota-prius-plug-in-hybrids-to-fleet/?_php=true&_type=blogs&_php=true&_type=blogs&_r=1
112. Zipcar I (2014) Zipcar and Vauxhall charge up electric vehicle sharing pilot in London (cited 4 Mar 2014). Available from: http://www.zipcar.com/press/releases/zipcar-launches-pilot-partnership-with-vauxhall-ev
113. McKercher B, Prideaux B (2011) Are tourism impacts low on personal environmental agendas? J Sustain Tourism 19(3):325–345
114. Lush (2014) What's wrong with Palm? (cited 4 Mar 2014). Available from: https://www.lushusa.com/on/demandware.store/Sites-Lush-Site/en_US/AboutUs-EthicalCampaign?start=0
115. Lush (2009) Naked (cited 4 Mar 2014). Available from: https://www.lushusa.com/on/demandware.store/Sites-Lush-Site/en_US/AboutUs-EthicalCampaign?start=7
116. Zhu Q, Sarkis J, Lai KH (2011) An institutional theoretic investigation on the links between internationalization of Chinese manufacturers and their environmental supply chain management. Resour Conserv Recycl 55(6):623–630
117. Lush (2014) Green tips: lush bottles and pots 2010 (cited 4 Mar 2014). Available from: http://www.lushusa.com/on/demandware.store/Sites-Lush-Site/en_US/Blog-Posts?cid=Article_Green-Bottles
118. Lush (2014) New ways to re-use your black pots (cited 4 Mar 2014). Available from: https://www.lushusa.com/New-Ways-to-Re-use-Your-Black-Pots/article_reuse-black-pots,en_US,pg.html
119. Company HPD (2014) Product return and recycling (cited 4 Mar 2014). Available from: http://www8.hp.com/us/en/hp-information/environment/product-recycling.html#.U19LYq21Z_1
120. Lam JS, Lai K (2014) Developing environmental sustainability by ANP-QFD approach: the case of shipping operations. J Clean Prod 1–10
121. Carnevalli JA, Miguel PC (2008) Review, analysis and classification of the literature on QFD-types of research, difficulties and benefits. Int J Prod Econ 114(2):737–754
122. Chan LK, Wu ML (2002) Quality function deployment: a literature review. Eur J Oper Res 143(3):463–497
123. Bereketli I, Genevois ME (2013) An integrated QFDE approach for identifying improvement strategies in sustainable product development. J Clean Prod 54:188–198
124. Lai KH, Wong CWY, Lun VYH (2014) The role of customer integration in extended producer responsibility: a study of Chinese export manufacturers. Int J Prod Econ 147 Part B: 284–293

125. Guide VDRJ, Jayaraman V, Linton J (2003) Building contingency planning for closed-loop supply chains with product recovery. J Oper Manag 21(3):259–279
126. Kassinis G, Vafeas N (2006) Stakeholder pressures and environmental performance. Acad Manag J 49(1):145–159
127. Wong CWY et al (2012) The roles of stakeholder support and procedure-oriented management on asset recovery. Int J Prod Econ 135(2):584–594
128. Agency USEP (2013) Sustainable materials management: sustainable consumption and production: European Union policy (cited 6 Mar 2014). Available from: http://www.epa.gov/oswer/international/factsheets/200810-sustainable-consumption-and-production.htm
129. Society NG (2013) Greendex (cited 6 Mar 2014). Available from: http://environment.nationalgeographic.com/environment/greendex/?rptregcta=reg_free_np&rptregcampaign=20131016_rw_membership_r1p_intl_dr_w#close-modal
130. Society NG (2012) Greendex 2012: consumer choice and the environment—a worldwide tracking survey, p 204
131. Leonidoua LC, Leonidou CN, Kvasova O (2010) Antecedents and outcomes of consumer environmentally friendly attitudes and behaviour. J Mark Manag 26(13–14):1319–1344
132. Development IIfS (2013) Who are the green consumers? (cited 6 Mar 2014). Available from: http://www.iisd.org/business/markets/green_who.aspx
133. Company TN (2012) The global, socially conscious consumer (cited 6 Mar 2014). Available from: http://www.nielsen.com/us/en/newswire/2012/the-global-socially-conscious-consumer.html
134. Peattie K (2001) Golden goose or wild goose? The hunt for the green consumer. Bus Strategy Environ 10(4):187–199
135. Joergens C (2006) Ethical fashion: myth or future trend? J Fashion Mark Manag 10(3):360–371
136. Company TN (2011) The green gap between environmental concerns and the cash register (cited 4 Mar 2014). Available from: http://www.nielsen.com/us/en/newswire/2011/the-green-gap-between-environmental-concerns-and-the-cash-register.html
137. Company TN (2013) Will a desire to protect the environment translate into action? (cited 9 Mar 2014). Available from: http://www.nielsen.com/us/en/newswire/2013/will-a-desire-to-protect-the-environment-translate-into-action-.html
138. Incorporated C (2010) Shades of the green consumer
139. Niinimäki K (2010) Eco-clothing, consumer identity and ideology. Sustain Dev 18(3):150–162
140. Keller KL (1993) Conceptualizing, measuring, and managing customer-based brand equity. J Mark 51(1):1–22
141. Beard ND (2008) The branding of ethical fashion and consumer: a luxury niche or massmarket reality? J Dress Body Cult 12(4):447–468
142. Engel JF, Blackwell RD, Minlard PW (1986) Consumer behavior. Dryden Press, New York
143. Chan T-Y, Wong CWY (2012) The consumption side of sustainable fashion supply chain: understanding fashion consumer eco-fashion consumption decision. J Fashion Mark Manag 16(2):193–215
144. Cronin JJ, Taylor SA (1992) Measuring service quality: a reexamination and extension. J Mark 56(3):55–68
145. Baker J et al (2002) The influence of multiple store environment cues on perceived merchandise value and patronage intentions. J Mark 66(2):120–141
146. Grewal D et al (2003) The effects of wait expectations and store atmosphere evaluations on patronage intentions in service-intensive retail stores. J Retailing 79(4):259–268
147. Carrigan M, Attalla A (2001) The myth of the ethical consumer—do ethics matter in purchase behaviour. J Cons Mark 18(7):560–578
148. Roberts JA, Bacon DR (1997) Exploring the subtle relationships between environmental concern and ecologically conscious consumer behavior. J Bus Res 40(1):79–89
149. Stock K (2014) The brilliant business model behind H&M's clothes recycling plan. Bloomberg Businessweek 2013 (cited 9 Mar 2014). Available from: http://www.businessweek.com/articles/2013-06-24/the-brilliant-business-model-behind-h-and-ms-clothes-recycling-plan

150. Calvin B, Lewis A (2005) Focus group on consumers' ethical beliefs. In: Harrison R, Newholm T, Shaw D (eds) The ethical consumer. Sage Publications Ltd, United Kingdom
151. Brooks R (2014) H&M converts donated clothes into a new denim collection (cited 12 Mar 2014). Available from: http://www.sustainablebrands.com/news_and_views/Waste_Not/Ross_Brooks/HM_Converts_Donated_Clothes_New_Denim_Collection
152. Roberts JA (1996) Green consumers in the 1990s: profile and implications for advertising. J Bus Res 36(3):217–231
153. Bennett G, Williams F (2011) Mainstream green: moving sustainability from niche to normal. Ogilvy & Mather, New York
154. Monroe KB (1973) Buyers' subjective perceptions of price. J Mark Res 10(1):70–80
155. Miller C (1992) Development and human needs. In: Ekins P, Max-leef M (eds) Real-life economics. Routledge, London, pp 197–214
156. William BD, Monroe KB, Grewal D (1991) Effects of price, brand, and store information on buyers' product evaluations. J Mark Res 28(3):307–319

Chapter 4
Organizational Capabilities

Abstract This chapter elaborates the environmental management in the aspect of organizational capabilities. First of all, internal integration is the decline of functional barriers and the facilitation of real-time information sharing across different organizational functions. In particular, internal environmental information integration is helpful to reinforce corporate environmental adaptability, and improve the effectiveness of business functions in response to new challenges such as changes in customer expectation and market trends. Internal integration helps firms to achieve better performance in delivery, production costs, product quality, and production flexibility. Besides, environmental innovativeness a firm's ability to develop new environmental management practices that can facilitate the production, assembling and distribution of a product or service while reduce negative environmental impacts; hence, environmental innovativeness enables a firm to sustain business growth, raise the return on investment, and improve environmental performance. Similar to environmental innovativeness, environmental adaptability is vital for firm to effectively lower its environmental impacts. Environmental adaptability, which is referred to the degree of flexibility and responsiveness in responding to new market demands and requirements, allows firms to modify their extant products and processes and launch new products with shorter lead times to cope with the fast changing market expectations and environmental regulations. These benefits will put the firms at the advantage of first-movers which would gain larger market shares and higher investment returns.

4.1 Internal Integration

Supply chain integration is multidimensional and can be divided into three constructs: internal, supplier, and customer integration [1]. As found in the BSR/GlobeScan State of Sustainable Business Survey 2013, integrating sustainability into core business functions remains the most important challenge [2]. The study also revealed that

© The Author(s) 2016
C.W.Y. Wong et al., *Environmental Management*,
SpringerBriefs in Applied Sciences and Technology,
DOI 10.1007/978-3-319-23681-0_4

convincing company management of the value of integrating sustainability and changing management mentality are the most common barriers, while other barriers include the difficulty in demonstrating the value of integrating sustainability, particularly when financial constraints are present and profit is prioritized over sustainability [2].

Internal integration is referred to the strategic system of cross-functioning and collective responsibility across functions [3]. Internal integration practices can diminish functional barriers and facilitate real-time information sharing across key functions of a firm [4]. As such, it enables different functions, such as product design, procurement, production, sales and distribution, collaborate to meet customer requirements at a low total system cost [5].

Internal environmental information integration is concerned with the coordination of internal processes to achieve corporate environmental goals. It is related to the intra-organizational systems infrastructure to facilitate timely, accurate and standardized data exchange across organizational functions [6]. The extent to which a firm forms its communication and information sharing infrastructure across intra-organizational functions to enable organizational efforts in environmental management constitute environmental information integration [7].

4.1.1 Internal Integration Example

The National Institute of Water and Atmospheric Research

The National Institute of Water and Atmospheric Research (NIWA), a Crown Research Institute of New Zealand, successfully engages the organization from the top down in environmental management efforts. NIWA has a central environmental manager responsible for establishing NIWA's sustainability objectives and required outcomes. At the regional level, the managers and scientific employees are involved within the Environmental Responsibility Committees to provide ideas and inputs for the central environmental manager. The Committees are also responsible for managing sustainable business practices at both operational and scientific levels, and developing new sustainability ideas for national rollout. NIWA scientists work on a collaborative basis across different functions internally as well as with other institutes. Up until 2011, NIWA has achieved various environmental targets, including energy, water, waste and greenhouse gas. For example, they cut the fuel usage of research vessels by 30 % through reducing cruising speeds by 2 knots [8].

The International Business Machines Corporation

The International Business Machines Corporation (IBM), an American multinational technology and consulting organization, heavily emphasizes on internal integration for successful environmental sustainability efforts. Regarding internal integration efforts for environmental sustainability, Wayne Balta, Vice President of Environmental Affairs and Product Safety at IBM, believes that "such an operation requires collaboration of technical professionals across business units and staff functions as well as innovative technologies, ideas, and know-how, underlined by a management system supported by top executives [9]." For example, the IBM energy strategy entails reducing consumption of energy in its IT and manufacturing operations, and conserving energy and improving energy efficiency in facility operations. Through internal integration across functions and effective business processes, IBM has achieved environmental sustainability successfully. IBM has earned the EPA-sponsored climate protection awards five times since 1998, which Balta refers to as the "power of a globally integrated enterprise and cross-organizational collaboration [9]."

Caesars Entertainment

Caesars Entertainment Corporation (Caesars), an American casino-entertainment company, highlights the importance of employee engagement when implementing environmental management practices. CodeGreen, Caesars' environmental sustainability strategy was rooted in employees. Before Caesars' formal implementation of a sustainability strategy, its employee started to take measures to improve efficiencies, reduce energy consumption and waste, and increase recycling in 2007. The high level of employee engagement in CodeGreen enabled Caesars to reduce electricity consumption per square foot of air-conditioned space by 20 % from 2007 to 2012. The company also reduced carbon emissions to the atmosphere by 12 % and diverted 24 % of waste away from landfills in 2012 [10]. Moreover, the successful employee engagement in CodeGreen is found to positively influence such key job satisfaction measures as pride, supervisor satisfaction, willingness to recommend, ethical leadership, loyalty, and discretionary effort. Such an involvement also contributes to a low human resources turnover rate of 16.1 %, which is below the industry average at around 22 % [10].

4.1.2 Internal Integration Impacts

A McKinsey survey finds that the share of respondents saying their companies' top reasons for addressing sustainability include improving operational efficiency and lowering costs increased from 14 % in 2010 to 33 % in 2011 [11]. The survey also finds that firms integrate environmental sustainability across functions by 67 % of firms integrating environmental sustainability issues into mission and values, 60 % into external communications, 59 % into corporate culture and 58 % into internal communications. The survey also found that the employees of firms that consider environmental sustainability a top-three priority of their business agenda are more knowledgeable about their companies' environmental sustainability activities [11], indicating internal integration for environmental information sharing. However, the relationship between sustainability and value creation remains unclear, and is a barrier towards successful internal integration towards sustainable performance.

The extant research suggests that internal integration is positively associated with delivery, production costs, product quality, and production flexibility.

According to the total quality management, product development, and production literature, internal integration is debatably the basis of supply chain integration, as it removes functional barriers [1], and enables cross-functional cooperation [5]. An aligned objectives and breaking down of barriers between organizational functions are needed for successful implementation of quality management [12]. The product development literature also suggests that internal integration facilitates product design, engineering, manufacturing, and marketing functions to work closely to enable simultaneous engineering and design for manufacturing [13]. The production management literature also suggests that internal integration allows the sharing of knowledge between functions and manufacturing plants [14], and improves coordination of production capacity to improve production flexibility [15] and delivery performance [16].

Prior research reveals that internal integration has positive impacts on production cost, production flexibility as well as delivery, and product quality [17]. Specifically, internal integration has greater impacts on product quality and production cost under a high environmental uncertainty, as these performance outcomes highly depend on internal functions instead of external collaboration with supply chain partners [18]. Further interviews with respondents in the study suggest that internal integration to reduce uncertainty by facilitating internal collaboration among functions to respond to high production cost and to reduce defect rate in production [19].

Corporate environmental adaptability, that is, when firms responding to new environmental requirements and needs while sustaining economic growth [20]. It is a vital environmental management capability to address future changes in the environmental market flexibly [21]. A recent study finds that internal environmental information integration contributes to corporate environmental adaptability, which improves the effectiveness of business functions in responding to new challenges, new market intelligence, such as new customer expectations and market trends [22]. Internal environmental information exchange enables communication and information sharing that supports cross-functional cooperation [23]. Corporate environmental adaptability refers to the ability of firms to support the restructuring and reengineering of business processes to be in line with new environmental regulations and market demands. It enables flexibility of business functions is improved, in turn increasing corporate environmental adaptability of a firm.

4.1.3 Summary

Internal, supplier, and customer integration constitute supply chain integration. Internal integration is the reduction of functional barriers and facilitation of real-time information sharing across internal functions. Internal environmental information integration is concerned with the organization of internal processes to achieve corporate environmental goals, it involves communication and information sharing infrastructure across a firm's functions to support organizational efforts in environmental management.

Internal environmental information integration is useful for increasing corporate environmental adaptability, and improving the effectiveness of business functions in responding to new challenges, new market changes, such as customer expectations and market trends. Managers should be aware that internal environmental information integration is important for firms to improve corporate environmental adaptability to address changing environmental demands and requirements. Especially for firms facing encountering uncertainties in coping with environmental management requirements in their marketplace, they should proactively develop internal environmental information exchange in order to increase flexibility.

Internal integration can achieve better performance in relation to delivery, production costs, product quality, and production flexibility. For improvement in non time-based performance and internally dependent performance outcomes, such as product quality and production cost under a high level of environmental uncertainty, managers should focus on internal integration. As performance outcomes heavily rely on internal cooperation and are less sensitive to external input and collaboration, improvement of product quality and production cost under a high environmental uncertainty can be achieved through successful internal integration.

4.2   Environmental Innovativeness

The environmental performance of organizations is now an important aspect for evaluation. Environmental management capability covers environmental innovativeness and environmental adaptability. A firm's environmental management capability refers to the ability to develop innovative processes and products as well as being adaptive to market needs at the same time [24]. Environmental innovativeness refers to a firm's ability on developing and adopting products and processes that contributes to sustainable development.

4.2.1   Definition of Environmental Innovativeness

Environmental innovativeness can be defined as having novel practices in the organization that facilitates the production, assembling and distribution of a product or service with reduced environmental risk, pollution and other negative impacts. Organizations implementing new changes focusing on the environment, which involves companies' products, manufacturing and marketing, is an act of environmental innovativeness. Whether it is big or small, radical or incremental, these changes to develop unprecedented ideas or to improve performance are part of environmental innovation as long as the goal is to decrease the firm's environmental impacts [25].

The organizational dynamic capabilities theory outlines that environmental management capabilities refers to the capabilities of firms to innovate and adapt to changing market needs [26] that allows firms to sustain business growth [27]. To innovate means that firms need to be proactive in developing new ways and to revolutionize processes and products for improvement [28]. Dynamic capabilities theory suggests that being quick in response and active in product innovation would allow firms to have improved performance, while having the capabilities to effectively coordinate and manage internal and external competencies [27].

According to the dynamic capabilities theory, corporate environmental innovativeness refers to a firm's ability to develop and adopt new environmental management practices that can improve the efficiency and effectiveness of its environmental protection efforts [29]. Proactive actions such as making new investments and seizing opportunities to develop environmentally friendly products and new ways of production to reduce pollution are examples of environmental innovation. Streamlining the supply chain and improving processes between a focal firm and its suppliers and customers is vital for corporate environmental innovativeness [30]. Because improving communications in the supply chain enables resource visibility and information sharing to facilitate innovation [31].

One proactive means through which a firm can promote its environmental innovativeness is environmental governance, which is increasingly requested by customers and other stakeholders. Environmental governance includes development of corporate environmental policies and commitments, and communications of information about organizational intention, resources and performance [32]. There are two opposing viewpoints on the economic and financial impacts of environmental governance. On one hand, environmental initiatives will raise the operations and pollution compliance costs of firms, thereby eroding their profitability. On the other hand, environmental governance enables firms to better utilize the resources, increase production efficiency and enhance adaption to the environments, which is associated with cost reduction and profitability improvement [33, 34]. Recent study addresses the influence of environmental governance on corporate reputation and customer requirement. It is found that environmental governance can contribute to corporate reputation and customer satisfaction in order to improve economic performance [32].

Tang et al. [32] showed that environmental governance delivers economic performance through two channels: first, environmental governance serves to signal the firm's green reputation which is positively associated with overall corporate reputation so that the economic performance can be bolstered; second, since environmental governance can better satisfy the customer needs, firms can increase their sales and achieve higher economic gains.

Environmental management becomes an increasingly important and hot topic for many industries probably because the eco-damaging problems caused by human activities, such as greenhouse gas emissions and solid waste generation, are widely acknowledged and concerned. In particular, "greening" the operations and services has been gaining priority in the management of logistics industry. Green logistics management is novel and unique because it emphasizes to the mitigation of environmental damages that generated during the whole product life-cycle. It is an approach that takes into account product return and recycling, eco-efficiency, and environmental management systems for dealing with the environmental risks and complying with the potential environment regulations in goods transport [35, 36]. Other than internal production activities such as product development and manufacturing, managing the flows of physical products is vital for environmental protection in logistics industry [37]. Based on empirical analysis about the Chinese manufacturing exporters, Lai and Wong [38] found that green logistics management can positively affect exporters' environmental and operational performance.

Environmental management has increased the awareness not only in logistics industry but also, to a broader extent, in supply chain management. From the perspective of diffusion of innovation, green supply chain management (GSCM) can be viewed as an advanced environmental innovation for firms to improve environmental performance [39]. GSCM can also be used together with other environmental innovations such as cleaner production and environmental management systems. Since GSCM considers multiple participants of a supply chain and the balances of costs and benefits between different participants, it is relatively technically complicated relative to other innovations. GSCM usually starts on a trial basis, and gradually extends to more complete and full adoption [39, 40]. By acting ahead of the peers, early adopters of GSCM can establish pioneer costs and services improvements so that they are able to achieve greater performance gains. Drawing on the theory of innovation diffusion, Zhu et al. [40] considered GSCM as an organizational technical innovation and showed that the advanced manufacturers are often the early adopters of GSCM.

4.2.2 Environmental Innovation Examples

Sustainability consulting firm Pure Strategies published a report in 2014 specifically investigating business benefits from product based on quantitative surveys of 100 leading global consumer product companies [41]. According to the report, 97 % of the "performing companies", those that have already realized widespread financial and organizational value from their sustainability efforts, believe that setting product sustainability goals allow them to improve financial and environmental outcomes.

Many firms involve environmental innovation in sustainable products development. A report done by analyst firm Verdantix finds that firms focusing on product sustainability would achieve business benefits. The report finds that among the group of firms interviewed, at lead 80 % achieved cost savings on logistics and materials, and 67 % mitigated resource scarcity and risks associated with regulations [42]. Rodolphe d'Arjuzon, Global Head of Research at Verdantix, stated "The positive results from product sustainability investments include improved consumer trust in the brand, reduced logistics costs and lower exposure to supply risks. Insights into how to achieve these benefits are now diffusing across sectors such as footwear, consumer electronics and packaging [43]."

4.2.3   Adidas

Adidas, a German multinational sportswear manufacturer, has launched the Adidas DryDye in 2012 [44]. The DryDye technology is a polyester fabric dyeing process that uses no water, 50 % fewer chemicals and 50 % less energy than the traditional fabric dyeing process [45]. During the process, dye is injected onto the fabric together with compressed carbon dioxide. When the dyeing cycle is over, the carbon dioxide gasifies, so that the dye condenses and separates from the gas. The clean carbon dioxide can be collected and reused in the next dyeing cycle. The company started using this novel technology on 50,000 t-shirts. Since then, more and more sports products incorporated the DryDye. By the end of 2013, Adidas has saved 50 million liters of water by using 2 million yards of DryDye fabric in production [46].

In 2012, the company launched the Element Soul shoe and in 2013 the Element Voyager shoe. Both shoes uses fewer parts (12 parts compared to 30 in an average running shoe, which equals to 500 g decreased waste per shoe), maximum pattern efficiency (the Element Voyager shoe down to 5 % waste), and recycled materials (the complete Element Voyager shoe is made with environmentally preferred materials) [45]. The above are the examples of environmental innovativeness from Adidas. Adidas is constantly developing new technologies to reduce the use of scarce resources and to reduce environmental impact in production. The new DryDye technology, which is one form of environmental innovation, is proven a successful practice to reduce use of water. And through innovative product development with the Element Soul and Element Voyager shoe, negative environmental impact is reduced.

4.2.4   Reckitt Benckiser

Reckitt Benckiser (RB) is a multinational fast moving consumer goods company (FMCG). RB one of the first of its kind to implement a full lifecycle carbon reduction program in 2007 with the aim to reduce its' products' total carbon footprint by 20 % per dose by 2020, however the company managed to achieve this in 2012 [47]. Moving forward, RB continued to strive towards environmental goals and launched the Sustainable Innovation calculator in 2013, which is a measurement mechanism to assess the impacts of ingredients, packaging, carbon and water.

The RB Sustainable Innovation Calculator is a novel monitoring system developed by the company that facilitates the production, assembling and distribution of a product with reduced environmental risk, pollution and other negative impacts. It includes and calculates the whole lifecycle impacts of the company's products, from ingredients to packaging. This system facilitates the development of products with low environmental impact. It is an online tool to assist development teams to choose more sustainable options. The company has assessed over 500 new product innovations with the tool in 2013. Products are assessed as more sustainable when it performs better on at least one environmental aspect, for example carbon, water, packaging and ingredients, and no worse on any others [48]. RB aims to have one third of its net revenue to come from products that are more sustainable while keeping the same level of performance through innovative assessment of products and better product designs.

4.2.5 Environmental Innovation Implications

Environmental information integration practices would influence a firm's environmental management capabilities, which would consequently affect the financial and environmental performance of organizations. Prior research finds that customer environmental information integration, one dimension of environmental information integration, would strengthen environmental innovation. Through effective information sharing and communication with customers, the ability of firms to innovate new environmental management products to solve environmental problems is improved [22]. Customer environmental information integration can provide new market information and knowledge that can be used for developing new sustainable products and practices. According to supply chain management theory, the significance of customer environmental information integration indicates that environmental innovation stems from market intelligence that is gained through sharing information and communicating with customers [49].

Acquiring timely market information and requirements are is important to a firm [50], as it strengthens corporate environmental innovativeness, which is part of environmental management capabilities. Research findings indicate that corporate environmental innovativeness is positively associated with a firms business and environmental performance [22]. This is in line with the dynamic capabilities theory that a firm's innovativeness is vital organizational capabilities that contributes to business and environmental performance. It supports a firm in such areas as environmental product development and sustainable practices [51], it is also vital for firms to gain additional market share, increasing returns on investment to sustain business.

4.2.6 Summary

Environmental innovativeness is part of a firm's environmental management capabilities. Environmental innovativeness is when organizations implement novel practices in that improve the production, assembling and distribution of a product or service with reduced environmental risk, pollution and other negative impacts. According to the dynamic capabilities theory, corporate environmental innovativeness refers to a firm's ability to develop and adopt new environmental management practices that can improve the efficiency and effectiveness of its environmental protection efforts, this allows them to sustain business growth and improve environmental performance. Examples of environmental innovation includes but not limited to environmental governance, green logistics, GSCM, and the development of new technologies.

Environmental information integration practices influences environmental management capabilities. Firms should focus on implementing environmental information integration practices and environmental innovativeness at the same time. This would consequently improve the financial and environmental performance of organizations. Research findings indicate that customer environmental information integration, one dimension of environmental information integration, would strengthen environmental innovation. Firms can acquire valuable customer and market information through an integrated sharing and communication system. Sustainability managers should strengthen the firm's customer environmental information integration to facilitate a faultless information flow, thus latest market information for product or process innovation can be acquired. Through novel products and practices, environmental innovativeness also plays a role in improving a firm's return on investment, profit and market share.

4.3   Environmental Adaptability

4.3.1   Definition of Environmental Adaptability

Environmental adaptability is concerned with the degree of responsiveness and flexibility of a firm in responding to new environmental requirements and needs while sustaining economic growth [20]. It reflects how a firm could flexibly support the restructuring and reengineering of business processes to comply with new environmental regulations and market demands, such as use of alternative energy [52]. In the context of supply chain management, adaptability represents firm's capability in coping with the changing market needs and expectations. To date, environmental adaptability is a critical environmental management capability of addressing future changes in environmental policy and market expectations [53].

Environmental adaptability maintains the agility in firms' environmental management practices to adjust to comply with new environmental regulation and to fulfill the growing expectation of environmental protection. Such a capability enables firms to timely respond to new environmental requirements and commercializes new environmentally friendly products, contributing to organizational triple bottom line [54]. Environmental adaptability is valuable for firms in adapting to market changes and gaining investment payback in a short time period [22].

Green advertising is a popular form of adaptability in promoting the environmental attributes of business operations to cater for environmentally conscious customers [55]. A firm's green advertising may include press releases, disclosure of corporate environmental information, and sponsorship by environmental interest group, which informs and reminds customers about the firm's environmental stance [56]. Firms that seek to capitalize on growing public concern over environmental threats can adapt well to the ever-changing context and feed off of consumers' changing attitudes and behaviors. By publicizing the environmental actions, firms are able to proactively meet the requirement of government regulation and gain customer's confidence. Successful green advertising can help firms to excel from competitors in building favorable environmental image. Wong et al. [57] found that sharing information related to a firm's environmental efforts will generate a positive environmental reputation, which tends to benefit financial performance [58].

4.3.2 Example of Environmental Adaptability

4.3.2.1 BASF

BASF, the world's largest chemical company headquartered in German, is alertly prepared for new environmental regulations and policy pertaining to chemical products, making sure its knowledge of chemical manufacturing and testing was up-to-date with the legislation. When the Registration, Evaluation and Authorization of Chemicals (REACH) chemicals directive entered into force in the European Union (E.U.) in 2007, BASF has been gearing up for the new regulations for more than three years, and soon announced two initiatives in response to REACH, including a self-commitment for assessing chemicals and a consultancy report to help other firms to quickly meet or exceed requirements set by REACH. BASF is responsible for ensuring that its products are safe when chemicals are used properly and do not pose a hazard to human being or the environment. BASF aims to conduct a review and risk evaluation of all its chemicals produced or sold worldwide in quantities exceeding one metric ton per year by 2015. BASF's Center of Competence for the Environment, Safety and Energy reported that it has registered approximately 2500 chemicals under the terms of REACH. BASF's REACH consultations gives manufacturers information about its chemicals falling under the new chemical regulations, and which products can be imported or sold in the E.U., helping many small and medium firms that often cannot afford comprehensive consulting around regulatory requirements build up extensive understanding about REACH.

4.3.2.2 Wal-Mart

Wal-Mart has long committed to natural resource conservation and environmental protection. Since the combustion of nonrenewable fossil fuels produces much greenhouse gas and other air pollutants, Wal-Mart recognizes that the development of inexhaustible and clean solar energy technologies can achieve great long-run benefits. The use of solar energy enhances Wal-Mart's energy security through alleviating the burden of combating climate change, lowering the emissions of air pollutants, promoting sustainability, and lowering cost of fossil fuels. In the US market, solar panel prices for commercial installations declined by 15.6 % from $4.64/W to $3.92/W in 2013. Solar power currently supplies 2 % of the electricity needs and is projected to grow to 16 % by 2020. Among all the US companies, Wal-Mart has the highest solar energy capacity with 65,000 kW of solar power, which is enough to supply the annual energy demand of more than ten thousand households. Wal-Mart recently installed 10 new solar rooftop systems in Maryland, totaling more than 13,000 panels. With more than 200 solar installations across the country, Wal-Mart plans to build another 1000 installations by 2020, with the ultimate goal of supplying its whole energy needs with solar and other renewable energy [59].

4.3.3   Environmental Adaptability and Environmental Information Integration

Environmental adaptability is closely related to environmental information integration (EII), a concept referred to a firm's capacity of sharing information on environmental management with its supply chain partners [7, 22, 60, 61, 62]. EII reflects a firm's connectivity to acquire and disseminate information to coordinate environmental management practices, ranging from eco-product design, asset recovery, reuse, and components recycling, in order to mitigate the environmental footprint throughout product life cycle [63]. From the perspective of coordination along the supply chain, EII involves internal and external environmental management practices [6], including internal EII, customer EII, and supplier EII.

Internal EII is concerned with the extent to which a firm configures its communication and information sharing infrastructure across functions within the firm to facilitate efforts in environmental management [23, 64]. As internal EII facilities cross-functional adjustments in practice, it enables firms to improve adaptability in responding to changing market demands, expectations, and regulatory requirements. For example, Dentsu Inc. a Japanese advertising and public relations company, fosters environment-related communication among its internal departments. One such effort is the creation of the ECONNECT in-house database to share knowledge and information. ECONNECT which contains project summaries and internal personnel data is publicly available among employees. It also incorporates an online magazine and other features to share environment-related topics and case studies of the latest environmental measures [www.dentsu.com].

Customer EII is concerned with firms' infrastructure for sharing environmental information with their customers and acquiring information on new needs and expectations of customers. Customer EII supports information sharing and communication with customers, thereby improving firms' responsiveness to new market demands and lowering uncertainties in eco-product design, asset recovery, and reverse logistics scheduling [1, 65]. Through customer EII, firms not only could inform and educate customers about alternative products and practices, but also could explore more opportunities of business and environmental management [66]. For example, Apple Computer has offered free take-back and recycling of old computers to consumers who purchase new Macs at Apple stores or through online store since 2006. This old computers collection and recycling program is developed in response to the consumer's concern about the short product lifespan of Apple's iPod products. Applying its capable public relation resources, Apple makes this program visible to customers on an on-going basis. This response to consumer's environmental expectations enables Apple to maintain its technical competence by having understanding of consumer usage patterns and market needs. Given its success in the programme, its rivals such as Dell and HP also launched similar recycling programs.

Supplier EII is referred to the infrastructure that enables information sharing between a firm and its suppliers to facilitate environmentally responsible supply chain operations such as materials selection and treatment of manufacturing by-products. Supplier EII provides firms with information and knowledge related to alternative choices of materials, designs for material waste, reduction and reuse of scrap, and retrieved components to reduce material consumption and disposal to landfill [67, 68]. It enhances the consistency of environmental objectives along supply chains as well as cross-firm collaboration in undertaking environmental management practices. By promoting a collaborative and mutual understanding on partners' responsibilities and functions in environmental management, supplier EII helps develop flexibility in environmental management to adapt and respond to changes in environmental requirements and demands. For example, Disney demanded for products that are both effective and environmental friendly, and encouraged its suppliers to develop innovative products that meet health and safety goals and be harmless to the environment [69]. Another example is Honda Motor Co. that tracks emission of its suppliers' production of components and vehicle assembly to monitor and control their environmental impacts throughout its supply chain [70].

Wong [22] examines how each of the three types of EII contributes to a firm's environmental adaptability. It is hypothesized that each EII will engender environmental adaptability. Using firm-level data collected in Hong Kong, the study found that environmental adaptability is significantly positive associated with internal and consumer EII, whereas supplier EII alone does not suffice to improve environmental adaptability. Firms are recommended to promote their customer EII to achieve a seamless information flow in order to improve their environmental management capability.

4.3.4 Summary

Together with environmental innovativeness, environmental adaptability is crucial for firm to effectively manage its environmental footprint. The degree of flexibility and responsiveness in responding to new market demands and requirements constitute a firm's environmental adaptability. A firm with high environmental adaptability is able to modify their existing products and processes, and launch new products with short lead times, bringing such benefits as being the first-movers with larger market shares and investment returns. In a dynamic and competitive environment, environmental adaptability is important for firms to be able to cope with the fast changing market expectations and environmental regulations. To facilitate firms in enhancing environmental adaptability, it is important to allocate resources to promote green advertising and develop inter-organizational communication and information sharing mechanism with customers.

References

1. Flynn BB, Huo B, Zhao X (2010) The impact of supply chain integration on performance: a contingency and configuration approach. J Oper Manag 28(1):58–71
2. BSR/GlobeScan (2013) The BSR/GlobeScan state of sustainable business survey. Cited 16 Mar 2014. Available from: http://www.bsr.org/reports/BSR_GlobeScan_Survey_2013.pdf
3. Follett MP (1993) Freedom of coordination: lectures in business organization, 1987 (originally published in 1949). Garland Publishing, New York
4. Wong CY et al (2007) The implications of information sharing on bullwhip effects in a toy supply chain. Int J Risk Assess Manag 7(1):4–18
5. Morash EA, Dröge C, Vickery S (1996) Boundary spanning interfaces between logistics, production, marketing and new product development. Int J Phys Distrib Logistics Manag 26 (8):43–62
6. Truman GE (2000) Integration in electronic exchange environments. J Manag Inf Syst 17 (1):209–244
7. Lai KH et al (2006) Antecedents and consequences of electronic product code adoption and its implications for supply chain management: a framework and propositions for future research. Marit Econ Logistics 8(4):311–330
8. Accountants TIoC (2012) Integrating sustainability into business practices: a case study approach. Cited 16 Mar 2014. Available from: http://www.charteredaccountants.com.au/Industry-Topics/Sustainability/Integrated-reporting/Integrated-reporting/Integrating-sustainability-into-business.aspx
9. Balta W (2014) Successful sustainability comes from within. Building a smarter planet. Cited 16 Mar 2014. Available from: http://asmarterplanet.com/blog/2014/03/sustainability-2.html
10. Migita G (2014) Four Strategies for successful employee engagement: lessons from Caesars CodeGreen. Cited 16 Mar 2014. Available from: http://www.corporateecoforum.com/four-strategies-successful-employee-engagement-lessons-caesars-codegreen/
11. Company M (2011) The business of sustainability: McKinsey global survey results. Cited 16 Mar 2014. Available from: http://www.mckinsey.com/insights/energy_resources_materials/the_business_of_sustainability_mckinsey_global_survey_results
12. Deming WE (1982) Quality, productivity, and competitive position. MIT Center for Advanced Engineering, Cambridge, MA
13. Crawford CM (1992) The hidden costs of accelerated product development. J Prod Innov Manage 9(3):188–199
14. Roth AV (1996) Achieving strategic agility through economies of knowledge. Strategy Leadersh 24(2):30–36
15. Sawhney R (2006) Interplay between uncertainty and flexibility across the value-chain: towards a transformation model of manufacturing flexibility. J Oper Manag 24(4):476–493

16. Droge C, Jayaram J, Vickery SK (2004) The effects of internal versus external integration practices on time-based performance and overcome firm performance. J Oper Manag 22 (6):557–573
17. Wong CY, Boon-Itt S, Wong CWY (2011) The contingency effects of environmental uncertainty on the relationship between supply chain integration and operational performance. J Oper Manag 29(6):604–615
18. Ragatz GL, Handfield RB, Petersen KJ (2002) Benefits associated with supplier integration into new product development under conditions of technology uncertainty. J Bus Res 55 (3):389–400
19. Lai K-H, Wong CWY, Cheng TCE (2008) A coordination-theoretic investigation of the impact of electronic integration on logistics performance. Inf Manag 45(1):10–20
20. Milne MJ, Kearins K, Walton S (2006) Creating adventures in wonderland: the journey metaphor and environmental sustainability. Organization 13(6):801–839
21. Colby ME (1991) Environmental management in development: the evolution of paradigms. Ecol Econ 3(3):193–213
22. Wong CWY (2013) Leveraging environmental information integration to enable environmental management capabilities and performance. J Supply Chain Manag 49 (2):114–136
23. Mukhopadhyay T, Kekre S, Kalathur S (1995) Business value of information technology: a study of electronic data interchange. MIS Q 19(2):137–156
24. Quak HJ, de Koster MBM (2007) Exploring retailers' sensitivity to local sustainability policies. J Oper Manag 25(6):1103–1122
25. Dias Angelo F, Jabbour CJC, Vasconcellos Galina S (2012) Environmental innovation: in search of a meaning. World J Entrepreneurship Manag Sustain Dev 8(2/3):113–121
26. Beske P (2012) Dynamic capabilities and sustainable supply chain management. Int J Phys Distrib Logistics Manag 42(4):372–387
27. Teece DJ, Pisano G, Shuen A (1997) Dynamic capabilities and strategic management. Strategy Manag J 18(7):509–533
28. Ramus CA (2001) Organizational support for employees: encouraging creative ideas for environmental sustainability. Calif Manag Rev 43(3):85–86
29. Fong C-M, Chang N-J (2012) The impact of green learning orientation on proactive environmental innovation capability and firm performance. Afr J Bus Manag 6(3):725–727
30. Lee H (2000) Creating value through supply chain integration. Supply Chain Manag Rev 4 (4):30–39
31. Geffen CA, Rothenberg S (2000) Suppliers and environmental innovation: the automotive paint process. Int J Oper Prod Manag 20(2):166–186
32. Tang AK, Lai K-H, Cheng TCE (2012) Environmental governance of enterprises and their economic upshot through corporate reputation and customer satisfaction. Bus Strategy Environ 21(6):401–411
33. Shaft TM, Sharfman MP, Swahn M (2002) Using interorganizational information systems to support environmental management efforts at ASG. J Ind Ecol 5:95–115
34. Uotila J, Keilt MM, Zahra SA (2009) Exploration, exploitation, and financial performance: analysis of S&P 500 corporations. Strategy Manag J 30(2):221–231
35. Sharma S, Henriques I (2005) Stakeholder influences on sustainability practices in the Canadian forest products industry. Strategy Manag J 26(2):159–180
36. Hitchens DMWN (1999) The implications for competitiveness of environmental regulations for peripheral regions in the EU. Omega 27(1):101–114
37. Gonzalez-Benito J, González-Benito O (2005) Environmental proactivity and business performance: an empirical analysis. Omega 33(1):1–15
38. Lai K-H, Wong WC (2012) Green logistics management and performance: some empirical evidence from Chinese manufacturing exporters. Omega 40:267–282
39. Narasimhan R, Carter JR (1998) Environmental supply chain management. Center for Advanced Purchasing Studies. Arizona

40. Zhu Qinghua, Sarkis J, Lai K-H (2012) Green supply chain management innovation diffusion and its relationship to organizational improvement: an ecological modernization perspective. J Eng Tech Manag 29:168–185
41. Strategies P (2014) The path to product sustainability: a pure strategies report. Glucester, MA
42. Verdantix (2013) Product sustainability: the dos, don'ts and business benefits, p 26
43. Elks J (2013) Report: product sustainability investments will accelerate as business benefits revealed. Sustainable Brands 3 Sept 2013. Cited 4 June 2014. Available from: http://www.sustainablebrands.com/news_and_views/communications/report-product-sustainability-investments-will-accelerate-business-ben
44. Group A (2012) Introducing Adidas Drydye .Cited 1 June 2014. Available from: http://www.adidas.com/com/goallin/articles/2012/07/adidas-drydye/
45. Group A (2014) Sustainability innovation. Cited 1 June 2014. Available from: http://www.adidas-group.com/en/sustainability/products/sustainability-innovation/#/adidas-drydye/
46. Hower M (2014) Adidas details progress on supplier audits, sustainable materials use in 2013. Sustainable Brands 16 Apr 2014. Cited 1 June 2014. Available from: http://www.sustainablebrands.com/news_and_views/supply_chain/mike_hower/adidas_supplier_audit_coverage_reached_75_2013
47. Fox N (2013) Reckitt Benckiser smashes product lifecycle targets. Cited 1 June 2014. Available from: http://www.theguardian.com/sustainable-business/reckitt-benckiser-smashes-lifecycle-targets
48. Plc RBG (2013) Sustainability report, p 52
49. Tate WL, Ellram LM, Kirchoff JF (2010) Corporate social responsbility reports: a thematic analysis related to supply chain management. J Supply Chain Manag 46(1):19–37
50. Frohlich MT, Westbrook R (2002) Demand chain management in manufacturing and services: web-based integration, drivers and performance. J Oper Manag 20(20):6
51. Handfield R, Sroufe R, Walton S (2005) Integrating environmental management and supply chain strategies. Bus Strategy Environ 14(1):1–19
52. Colby ME (1991) Environmental management in development. The evolution of paradigms. Ecol Econ 3(3):193–213
53. Lee HL (2004) The triple-A supply chain. Harvard Bus Rev 82(10):102–112
54. Carter CR, Dresner M (2001) Environmental purchasing and supply management: cross functional development of grounded theory. J Supply Chain Manag 37(3):12–27
55. Kilbourne WE (1995) Green advertising: salvation or oxymoron? J Advertising 24(2):7–19
56. Narayanan S, Manchanda P (2009) Heterogeneous learning and the targeting of marketing communication for new products. Mark Sci 28:424–441
57. Wong CWY, Lai KH, Cheng TCE, Lun VYH (2012) The roles of stakeholder support and procedure-oriented management on asset recovery. Int J Prod Econ 135:584–594
58. Wong WYC, Lai KH, Shang KC, Lu CS (2014) Uncovering the value of green advertising for environmental management practices. Bus Strategy Environ 23:117–132
59. Finnigan J (2013) How utilities can adapt when big-box retailers go solar. Available from: http://www.greenbiz.com/blog/2013/08/16/utilities-adapt-big-box-solar
60. Almotairi B, Lumsden K (2009) Port logistics play form integration in supply chain management. Int J Shipping Transp Logistics 1(2):194–210
61. Bajwa DS, Lewis LF, Pervan G, Lai VS, Munkvold BE, Schwabe G (2008) Factors in the global assimilation of collaborative information technologies: an exploratory investigation in five regions. J Manag Inf Syst 25(1):131–165
62. Grover V, Saeed KA (2007) The impact of product, market, and relationship characteristics on inter-organizational system integration in manufacturer-supplier dyads. J Manag Inf Syst 23(4):185–216
63. Elliot S (2011) Trans-disciplinary perspectives on environmental sustainability: a resource base and framework for IT-enabled business transformation. MIS Q 35(1):197–236
64. Lai K-H, Wong CWY, Cheng TCE (2008) A coordination-theoretic investigation of the impact of electronic integration on logistics performance. Inf Manag 45(1):10–20

65. Ettle JE, Reza E (1992) Organizational integration and process innovation. Acad Manag J 34:795–827
66. Priem RL, Swink M (2012) A demand-side perspective on supply chain management. J Supply Chain Manag 48(2):7–13
67. Carter CR, Carter JR (1998) Interorganizational determinants of environmental purchasing: initial evidence from the consumer products industries. Decis Sci 29(3):659–684
68. Pagell M, Wu Z, Wasserman ME (2010) Thinking differently about purchasing portfolios: an assessment of sustainable sourcing. J Supply Chain Manag 46(1):57–73
69. Liroff R (2009) When the "FDA Approves Salmonella" how should your company respond? Available from: http://www.greenbiz.com/blog/2009/03/16/when-fda-approves-salmonella-how-should-your-company-respond
70. Bardelline J (2011) Honda takes green purchasing guide global. Greenbiz com. Available at http://www.greenbiz.com/news/2011/01/14/honda-takes-green-purchasing-guide-global

Chapter 5
Closed Loop Supply Chain

Abstract This chapter gives the reader an understanding of the managerial aspects of the reverse flows with various kinds of returns in manufacturing firms. With the increasing streams of returns over the product life cycle, including repair and replacement, end-of-use return, end-of-life return, production return flows, and distribution return, effective disposition of the returned products, components and packaging has grown as an urgent concern of manufacturing firms. Reverse logistics is intrinsically consistent with environmental sustainability. Manufacturing firms can reduce the emissions of hazardous emissions and energy consumption by maximizing tons per mile, consolidating shipments, optimizing the waste management and material recovery processes. These practices not only lower waste and pollution to protect the environment, but also enhance firms' profitability, asset utilization, and sustainable development. The other important topic in closed loop supply chain management is the asset recovery—the process of maximizing the value of unused or end-of-life assets through effective identification, redeployment, and divestment. Asset recovery enables businesses to capture the full value of returned products and save economic costs in remanufacturing and reuse without violating the regulatory requirements, thereby strengthening the cost reduction and product quality control.

5.1 Reverse Logistics

5.1.1 Definition of Reverse Logistics

The conventional logistics is referred to as the movement of goods throughout the supply chain network from production to distributor or customer. In contrast, reverse logistics represents movements of products from the point of consumption to the point of origin. Reverse logistics management is concerned with the management approach that facilitates firms to be more environmentally-efficient through

© The Author(s) 2016
C.W.Y. Wong et al., *Environmental Management*,
SpringerBriefs in Applied Sciences and Technology,
DOI 10.1007/978-3-319-23681-0_5

recovering, recycling, and reusing product components and parts [1]. Reverse logistics has been regarded as one of the 21 top warehousing trends in the 21st century [2].

Instead of carting products to landfills, firms are encouraged to recover the residual value of the products through product take-back, remanufacturing, recycling and so forth, with aim to reduce use of virgin materials for new product development. In the European Union, the WEEE (Waste Electrical and Electronic Equipment) and RoHS (Restriction of Hazardous Substances) restrict the use of hazardous substances in electrical and electronic equipment and require firms to collect, recycle, and reuse parts and components. Similar rules are being developed in the United States. For example, California enacted the 2003 Electronic Waste Recycling Act (EWRA) that followed the EU RoHS directive. Japan's recycling laws also spur the domestic manufacturers to perform in accordance with RoHS guidelines (http://www.meti.go.jp/policy/recycle/main/english/law/promotion.html).

Zappos.com is an online retailer of shoes, apparel, and home goods. As part of its environmental management initiatives, Zappos.com encourages its customers to order any styles, colors, and sizes of products, try them on, and return for free if the products do not fit [3]. Such a return policy allows Zappos.com to increase customer satisfaction, while reduce waste of products. Customers can simply use online UPS Returns® and call for pickup. The seamless reverse logistics experience offered by UPS leads to a 75 % repeat consumer rate. Not only can shoppers easily return their purchases, Zappos could also monitor the incoming returns and inbound goods from vendors, enabling them expedite to replenish stock for resale, reducing inventory in the supply chain.

Another example is Haier Group, which is one of the world's major consumer electronics and home appliances enterprises founded in Qingdao City, China. Haier is proactive in various environmental management initiatives. With support from the local government, Haier launched a project to handle its end-of-life products by establishing a site with an annual capacity to process 200,000 used electronic products [4]. In 2007, Haier invested a total of RMB 80 million (approx. US$12 million) to construct a recycling center in Qingdao. Yang et al. [5] estimated that the center could recycle about 200,000 used home appliances, including air conditioners, TV sets, and washing machines every year.

5.1.2 Types of Reverse Logistics

Any process or management after the sale of the product involves reverse logistics. The returned products can result from damaged in shipments, incorrect delivery, overstock, and return due to customer dissatisfaction. Flapper et al. [6] suggests that the types of reverse logistics can be categorized into the following four types.

- *Production-based*: Production-based of reverse logistics is referred to the handling of obsolete materials, production scraps, and product defects. Manufacturers would organize shipping and testing the returned product, and then dismantling, repairing, recycling, or disposing of customer returned defective purchase. The returned product would hence travel backwards through a supply chain in order to facilitate firms to recapture the remaining values of the returned products.

- *Distribution-based*: Distribution based of reverse logistics is considered the returns of products due to products are sold with a return option, wrong deliveries, or non-conformance to product specifications. For instance, a distributor may reject the order of Christmas trees if the order is delivered months before Christmas time. Distribution based reverse logistics may also due to safety concern. For example, product recall during distribution in the food and vehicle industries, where products are found with issues that may affect people health and safety. For example, in March 2012, a German luxury carmaker, BMW, recalled about 1.3 million cars worldwide because of battery malfunctioning is found with potential of charring or fire in extreme cases [link: http://www.bbc.com/news/business-17514172].
- *Use-related*: Use-related reverse logistics is concerned with products maintenance and after-sales services. For example, there are a large number of devices and components on airline fleets that need the daily maintenance.
- *End-of-Life*: End-of-life reverse logistics is related to products return to distributors or manufacturers due to end-of-life product that cannot longer be used or repaired in its current form. Yet, the components and materials of the products can be recycled and reused to make products that can be placed in a forward supply chain. For example, printer ink cartridges can be returned to their manufacturers to refill the ink reservoir to resell to the consumers. This closed-loop supply chain of ink cartridges is highly efficient as almost no energy is wasted on melting and recreating the cartridges.

5.1.3 Major Practices of Reverse Logistics

For a firm, adopting reverse logistics is an attempt to explore innovative ways to protect the environment. A growing number of manufacturers have integrated the practices of reverse logistics in their operations to develop sustainable competitive advantage.

Literature has different views in identifying and classifying the practices of reverse logistics. Carter and Ellram [1] proposed a pyramid form of practices: resource reduction is the most fundamental goal, followed by reuse maximization, recycling, and lastly, disposal. Rogers and Tibben-Lembke [7] listed five common ways of dealing with returned products—resold "as is", remanufacturing or refurbishing, recycling or landfill, repackaging, recovering primary materials. Stock [8] noted that reverse logistics combines multiple activities such as refurbishing, recycling, repair, and disposal. Prahinski and Kocabasoglu [9] summarized four disposition alternatives—waste management (i.e. incineration and landfilling), materials recovery (i.e. recycling and cannibalization), reuse/resell, and product upgrade (i.e. repackaging, repair, refurbish or remanufacturing). More recently, Lai et al. [10] suggested that there are six major aspects of practicing reverse logistics, each of which is elaborated as follows.

- Waste Management
  Mass production and consumption has made waste a serious environmental concern. Any product that is discarded because of perceived useless or unneeded during its life cycle will produce wastes. Waste management practices include incineration and landfilling. Heidrich et al. [11] suggested that waste management has been adopted by firms in order to facilitate their materials flows and to meet the requirements of stakeholders. Sarkis and Dijkshoorn [12] conducted a survey on small-and medium-sized manufacturing enterprises and found that solid waste management benefits their performance. Wal-Mart, the largest American multinational retailer, processes 45 million cases of returned merchandises per year. Inside each massive warehouse, Wal-Mart sorts the returned merchandises into four categories, namely vendor for credit, donation, recycle, and landfill. As part of Wal-Mart's sustainability plan, items such as perishable foods should go into the organic compost pile for organic farming [13]. On the other hand, Nike works with its materials suppliers to recycle factory wastes for new footwear manufacturing. Such practice enables Nike to cut down its waste posed to landfill from 25 % in 2007 to 13 % in 2009, while benefit from reduced purchase and consumption of new raw materials.

- Recycling
  Recycling is a process of collecting the used materials that can be disassembled and changed into new products. This practice helps to better utilize the scarce virgin raw materials and reduce the wastes resulting from disposal of used products. Electronic components are often collected by their manufacturers, e.g., HPs, Apples, and Dells, where they could be incorporated into the next generation of products. The recycling of copper in support of sustainable development and the environmental protection is another example on the value of recycling [14]. Take Fuji Xerox, a joint venture partnership between the Japanese photographic firm Fuji Photo Film Co. and the American document management company, for example. The company provided a list of various recycling projects in their clean factory goals. It recycled waste toner as binder to extract iron from steel manufacturing dust, and recycled waste plastic by turning it into chips at a raw material maker [2]. Another real-life example is that DHL Envirosolutions, a newly-developed subsidiary of DHL mail and freight delivery in 2011. The subsidiary offers trash and recycling hauling to pub chain, JD Wetherspoon, which has 800 pubs and all of them sorting waste at their locations. With the aid of DHL new system, JD Wetherspoon increased the recycled amount of aluminum by 89 %, steel by 57 %, and cardboard and paper by 19 %. As a result, the company saved £150,000 (Approx. US$243,000) in landfill fees.

- Reuse

  Reuse is referred to collecting products or packaging materials to be cleaned and reused. It includes conventional reuse where the item is used again for the same function and new-life reuse where it is used for a different function. By taking useful products and exchanging them without reprocessing, reuse help save resources, energy, and time. One example is the reuse of secondary packaging materials that are used for shipment purpose. The reuse of packaging materials can benefit manufactures since the materials are used repeatedly without requiring additional resource consumption. Thomas [15] found that an increase reuse of products can reduce consumption of new materials and products while reducing costs for consumers. Lenovo contributes in product reuse by offering asset recovery services and product take-back recycling programs. In 2011, Lenovo processed 700 metric tons of customer returned computer equipment, most of which were reused as products or parts, recycled as materials, incinerated with waste to energy recovery. FareShare (www.fareshare.org.uk) is a national organization collaborating with over 100 food businesses, wholesalers, and retailers to reduce food waste by redistributing fresh food in surplus to day centers and night shelters for homeless people. The scheme started operations since 2004 and currently owns 17 centers around the UK. From 2010 to 2011, the food collected contributed to over 8.6 million meals and around 35,500 people in-need every day [16]. Furniture Reuse Network (FRN) (www.frn.org. uk) is a nationwide coordinating body for furniture and appliance reuse and recycling organizations in the UK. The FRN promotes the reuse of unwanted furniture and household appliance, with two million items per year being reused and delivered on to low-income families. As a result, over 300,000 fridges are collected annually, and around 90,000 tons of waste is diverted from landfill through the FRN network [16].

- Reprocessing

  The returned products handled by manufacturers can be in the form of excessive stocks, end-of-season items, unsold or damaged goods, recalls, returns, and packaging materials. Where the origin of the material is scrap or waste, the recovered end-product can be referred as "reprocessed" which distinguishes from the "recycled" material is derived from the genuine pose-use products [2]. Reprocessing can help to recover potential value of the returned products. Manufacturers refurbish, repair, and reprocess returned products by selling them to discounters or using them for spare parts. For example, Coca Cola Enterprise, the world's third-largest independent Coca-Cola bottler, and a recycling firm ECO Plastics announced in March 2011 to jointly build a new plastic recycling plant in the UK. The construction of the plant is capable of annually reprocessing 75,000 metric tons of polyethylene terephthalate bottles into high-grade recycled plastic suitable for beverage and food packaging. In China, ZTE Corporation ensures the safe handling, movement, storage, use, recycling, and disposal of hazardous returned products if they pose harms to the environment. Other than the consideration for customer satisfaction, reprocessing of returned products to specific quality would help manufactures to establish an environmentally friendly image to the public with an environmental strategy to minimize waste. Ascent Healthcare Solutions, an independent third-party reprocessor of single-use medical devices in the United States, is recognized for its contribution to the healthcare industry by addressing environmental issues through the reduction of medical waste. It is expected that Ascent has eliminated over 7000 tons from U.S. landfills in 2006.

- Materials Recovery
  Product recovery is a useful practice in a closed-loop supply chain through which products are collected, reprocessed, and redistributed to the customers [17]. It aims to reflect both the economic and the ecological value of used products for eliminating wastes prior to disposal. For example, Tianzhong Electromechanical Company, the largest copier remanufacturer in Asia, collects used copiers for remanufacturing and sells them to emerging markets. This practice reduces new resources consumption for producing copiers, extends the lifespan of electronic goods, and benefits the company's environmental and economic performance. The rising number of vehicles makes the disposal of used tires an urgent environmental concern. Recovery of used tires is an example that illustrates the value of product recovery in reducing resources consumption and damages caused to the environment [18]. Recovering materials from returned products means lower frequency of product disposal and hence higher economic benefits for firms and social benefits for the environment. Another example is the recovery of materials (e.g., copper) from discarded electric products such as printed circuit board (PCB) and air conditioners by Huaxin Green Spring Environmental Protection Company in Beijing. Material recovers provide opportunities for manufacturers to reduce costs in material sourcing, improve corporate reputation, and satisfy the escalating customer request on environmental protection.

- Eco-design
  Eco-design is an approach that takes into account environmental influences in the product design and development, which is a major aspect of the prevention and reduction at the source of environmental impacts [19]. Playing a significant role in a firm's environmental management programs, product design encourages the use of standardized materials and adoption of modular design with aim to reduce waste in production. The first wave of eco-design dates back to the early 1970s and it aims at reducing the quantity of solid wastes ended up in landfills [19]. Eco-design is associated with the approach associated with a close loop of input material flows. At the design stage, the product-related environmental damage can be mitigated to design products with components that are easy to be disassembled for reuse and recycling to capture the residual values of returned products for the use of standardized materials and modular design. For example, TCL, an electronic home appliance manufacturer in China, pay special attention to energy consumption, health, and safety problems. All of TCL color televisions are designed to be energy-saving with certification consuming less than 3 W on

sleeping mode. Recently, TCL employs the optical injection technology with high-gloss materials to control spray and other highly polluting processes, thereby attaining zero-emission of wastewater, gas and oil residual per output value [10]. It also uses advanced disassembling technologies to transform used home appliances into renewable resources to build a circular economy-based supply chain. In short, design for reverse logistics addresses both product functionality and the minimization of life-cycle environmental impact.

Lai et al. [10] further use survey data to examine the contributions of the six aspects to the environmental performance of Chinese export-oriented manufacturers. It is found that only practicing waste management cannot improve a manufacturer's environmental and financial performance, which suggests that waste management is the final resort of reverse logistics. Supporting evidences are presented that recycling, reuse, material recovery, reprocessing, and eco-design could improve the environmental performance of the manufacturers. Overall, a manufacturer may hurt its image by implementing the rudimentary waste management only; rather, reverse logistics management has to be integrated with recycle, reuse, recovery, reprocess, and design [10]. From the perspective of product life cycle assessment, reverse logistics is expected to generate an environmentally beneficial impact [20].

5.1.4 Implementation of Reverse Logistics

Nowadays there are plenty of emerging eco-industrial parks built around the world following the principle of waste resource reduction and recycling [21]. However, Industrial plants tend not to commit into the eco-efficiency improvement or renewable resource substitution, if the associated costs are too high. Industrial waste reuse (IWR) is a feasible way of lowering carbon emissions without aggravating economic burdens on individual plant. For example, Zhang et al. (forthcoming) examines China's province-level IWR from 1999 to 2001 and identify that IWR will generate two effects on carbon emissions. First, IWR directly saves the materials for production and fully utilizes the disposal, which decreases the carbon emissions. Second, when a region's economic growth reaches a given level, IWR will contributes to carbon emission reduction with economic development being as a mediator [20].

While reverse logistics tends to lead to environmental gains, the adoption and implementation of the practices of reverse logistics still face some hurdles to overcome, especially in developing countries where environmental awareness is low. For example, it is recognized the Chinese firms has difficulties in processing the remanufacturing. Zhu et al. [22] conducted an empirical study to identify the barriers for China's truck-engine manufacturing from the perspective of reprocessing. They present convincing evidence that the key barriers are the lack of strong financial supports for China's remanufacturing technologies and equipment updates and innovations; other major barriers include lack of quality standards of reprocessed products, limited availability of old truck engines, and supporting marketing for remanufactured engines.

5.1.5 Summary

Nowadays, effective disposition of the returned products, components and pack-aging has been an increasing concern for firms. Prahinski and Kocabasoglu [9] suggests that the importance of reverse logistics for five reasons. First, product returns may reach a very high volume; in some industries, returns need handling exceeds 50 % of sales. Second, firms may see the promising potential to earn from the sales in secondary markets from the discarded products. Third, innovative alternatives such as repackaging, recycling, and reprocessing become more sus-tainable than high-cost inefficient landfill. Fourth, consumers may require the enterprise responsibility to take back products containing hazardous wastes. Lastly, the developed world like the European Union and the United States continue to press their enterprises to effectively manage their entire life of the product.

Reverse logistics is intrinsically consistent with environmental sustainability. It can be classified as production-based, distribution-based, use-related, and end-of-life. The common practices for reverse logistics include waste management, recycling, reuse, material recovery, reprocessing, and eco-design. By maximizing tons per mile, consolidating shipments, optimizing the waste management and material recovery processes, firms mitigate harmful emissions and energy usage with great potential to improve their profitability and asset utilization. Reverse logistics will not only directly reduces waste and pollution, but also benefits the environmental through the dual gain from economic development.

5.2 Asset Recovery

5.2.1 Definition of Asset Recovery

Asset recovery is referred to as the process of separating and inspecting returned products and components for remanufacturing and reuse to maximize the value of returned products [23]. It is regarded as a key management practice through which firms can recognize additional profits from products that are deemed inactive or of little monetary value [24, 25]. Proper separation and inspection of returned products into recovery are key activities of asset recovery that generate economic values and save the cost and time of redistributing products in the forward supply chain.

5.2.2 Asset Recovery in Green Logistics Management

Nowadays a growing number of firms are committed to asset recovery to satisfy customers who often expect firms to be environmentally responsible in their operations and product offerings. Asset recovery has its strength. It creates customer value by providing alternatives that are more environmentally responsible, yet cheaper, for consumption. Specifically, the refurbished and remanufactured products reduce the costs of product development. These products provide customers with a less expensive choice in the marketplace, while the recovered parts and components offer a source of spare parts and convenience for repair that enables customers to maintain their purchases at a lower expense instead of buying a new product [26].

Asset recovery contributes to greening the logistics chain through recycling materials, remanufacturing, and redistributing products, where its implementation is guided by firms' environmental management policies and systems [23]. Being an important process in green logistics management, asset recovery ensures that the returned products are effectively sorted and processed for remanufacturing and redistribution. Firms adopt asset recovery to retrieve the residual value of returned products in various environment management practices [27, 28].

First, asset recovery concerns the environmental management of industrial firms by minimizing waste and disposal to landfill. An exemplar of implementing asset recovery is Procter & Gamble Co. (P&G), an American multinational company produces pet foods, cleaning agents, and personal care products. Solid waste may be generated during production, use, and after-use of P&G products, which represent resource wastes and potential adverse environmental impacts. Noting such a significant environmental impact, P&G achieves zero manufacturing waste to landfill at 50 subsidiary sites worldwide by recycling, repurposing, and converting waste into energy. P&G's Global Asset Recovery Purchases (GARP) team is responsible for searching and collaborating with external partners who can turn something thrown away into something with value. For example, waste generated from P&G Charmin plant in Mexico is converted to make roof tiles for the local community. Similarly, scraps from its U.S. Pampers manufacturing plant are turned into upholstery filling, while waste from producing Gillette shaving foam at a U.K. site is composted and recovered into turf for commercial use. Since 2009, GARP created over US$1 billion in value from waste management for P&G.

Second, product design is an important factor to facilitate firms to plan for returned product inspection and disassembly [29]. Taking old garments recycling as an example, the garments that are made of a combination of fiber, fabric, button, and zipper need demerging and component categorization after returned for recovery. Missing or faded labels on the returned garments often prolong the inspection time for determining the materials used and viable recovery options [30]. Reverse engineering, a channel of helping firms understand the effective disassembly of products in asset recovery, can be costly and time consuming if the disassembly is not contemplated in product design [31]. Thus, firms could ease and expedite the recycling process by taking account of the possible missing and fading of labels in the garment design phase. Specifically, the different components used in making garments should be recyclable with minimal processing and labels should be placed properly to avoid easy detachment and fading.

Returned products can be recovered in a number of ways, including resold in their current conditions, remanufactured, and components and materials recovery [24].

Resalable products are usually the untapped product moving back through the supply chain probably due to incorrect delivery or overstock. They can be easily redistributed to the forward supply chain without further processing. Products need remanufacturing or refurbishing products can be distributed to alternative markets or aftermarkets. These two ways benefit manufacturers by reducing their efforts in processing returned product by disassembly and recycling of parts and materials [25]. Component and materials recovery is concerned with retrieving reusable components and materials for reprocessing, recycling, and cannibalizing. The products or components for which all three recovery ways are not feasible will be disposed to landfill, although disposal is the least preferable in environmental management. Overall, asset recovery increases product speed-to-market, facilitates the liquidation of used, obsolete or defective inventory, and channels the returned products or components into new profit streams [32, 33].

An example of successful asset recovery implementation is International Business Machines Corporation (IBM). IBM is an American multinational technology and consulting corporation that manufactures and markets computer hardware and software, and offers infrastructure and consulting services of mainframe computers and nanotechnology. IBM's Global Asset Recovery Services (GARS) is one of business lines in IBM. It manages reusing internal assets, remarketing end-of-lease IBM system assets, and offering an end-of-life green management for the disposal of scrap information-technology equipment. During the period of 2011–2013, GARS achieved good progress and established leadership in environmental affairs in the industry. GARS enables IBM to replace and consolidate older technology hardware with more energy-efficient refurbished systems to save energy. Moreover, GARS achieved excellent waste minimization and pollution prevention with less than 0.7 % of materials sent for scrap and disassembly was incinerated or landfilled. More than 90 % of assets sent for refurbishment were eventually resold and reused with the amount of US$761 million was gained from the reuse of 1293 IBM System z® and IBM Power Systems™ equipment.

IBM's counterpart, Lenovo Group Ltd. (Lenovo), a Chinese computer technology company, has processed more than 107,800 metric tons of returned computer equipment through the contracted service suppliers since its establishment as a multinational company in 2005. Lenovo financed or managed the processing of more than 13,100 metric tons of Lenovo-owned and customer-returned computer equipment in 2012. About 9.8 % of the returns were reused as products or parts, 84 % was recycled, 3.2 % was incinerated for energy recovery, and 3 % was disposed by landfill. The recycled customer returns in 2012 accounts for 7 % of the total weight of new products launched to the market in 2008. Over 80 % of all Lenovo supplier processed end-of-life computer equipment is recovered materials (e.g., precious metals, ferrous and non-ferrous metals, plastics, glass) which are in turns used to produce new IT or non-IT products worldwide. Since May 2005, Lenovo suppliers have processed over 183 million pounds of Lenovo owned and customer returned equipment producing a large amount of reusable materials. In addition, Lenovo provides its large-enterprise customers with asset recovery services, including providing PC take-back, refurbishment, and recycling, so as technology equipment maintenance. Lenovo offers additional services (e.g. inventory control, value assessment of equipment, and on-site uninstallation) that provide its business customers a step-by-step solution to manage the life-cycle of equipment. Lenovo's expertise in managing the end-of-life computers saves customers' time and money spent on the disposal of their surplus or replaced equipment.

5.2.3 Asset Recovery, Financial Performance, and Product Quality

While existing studies on asset recovery are raised surrounding product eco-design for remanufacturing [34], product collection arrangements [35, 36]; recycling, and reuse [37], Wong et al. [38] shed some light by investigating the effects of asset recovery on financial performance and product quality—two crucial measures of environmental management performance. The degree of asset recovery implementation is positively correlated to the financial performance of manufacturers due to three reasons [38]. First, asset recovery reduces the firms' demand for new materials to manufacture new products, thereby increases saving in material costs and decreases unit cost. Second, making use of the retuned products and components reduces investment in the inventories of new products and components. Third, resale of the remanufactured products and refurbished components to aftermarkets increase manufacturers' revenue [26]. On the other hand, asset recovery benefits the product quality of manufacturers. Product quality can be measured by advancement in manufacturing that shortens the lead-time of product development and elevates the product acceptance in the market [39]. Asset recovery involves testing and inspecting, and disassembling returned products [24], and therefore presents valuable feedback for firms to have in-depth insights and experience on product functions, design, durability, and performance.

Wong et al. [38] argued that the positive relationship between asset recovery and financial performance/product quality are likely to be strengthened if manufacturers possess a high level of stakeholder support or procedure-based management practices. On one hand, stakeholder group, consisting of customers, suppliers, regulatory, and staff, can exert pressure that influences organizational decisions on environmental protection such as asset recovery. Specifically, customers return unused, end-of-use, or end-of-life products via reverse supply chain, triggering asset recovery of manufacturers. Suppliers may collaborate with manufacturers to refurbish the returned components, while ensuring the quality of recovered components. Environmental policy guides firms to adhere to the predefined environmental objectives and management principles such as cost cut-down and product quality improvement [40]. Staff engages in asset recovery so as to ensure that asset recovery processes are consistently carried out to achieve high quality and low cost [41]. On the other hand, procedure-based management practices, which are characterized by formality, policies, and green logistics management, are able to enhance asset recovery by facilitating inspection, segregation, disassembly, and conversion of the returned products into resalable forms (e.g., refurbished products or recycled materials).

Using empirical data collected from Chinese export-oriented manufacturers. Wong et al. [38] show that asset recovery is positively associated with both the cost reduction and quality improvement of manufacturers. Stakeholder support promotes the capability of asset recovery to reduce the cost and to improve the product quality. However, the product cost and quality of Chinese manufacturers improve when their asset recovery is leveraged with by a low-level procedure-based management. The procedure-based management practices require strict guidelines to control processes. While the implementation of asset recovery confronts with too much uncertainty on volume, timing, quality of the returned products, the imposition of strict guidelines adversely affects the flexibility of manufacturers in adjusting their asset recovery.

5.2.4   Summary

Although returned products that are unused, end-of-use, or end-of-life generate reverse flows in a supply chain that can be costly for firms, asset recovery helps firms recapture the value of returned products, while complying with regulatory requirements. Firms can achieve double benefits through asset recovery: (1) reap the full values from the returned products and save economic costs in remanufacturing, and (2) reuse and recycle the resources to comply with environmental regulation and maintain the sustainability. Asset recovery plays a salient role in environmental management by minimizing waste and disposal to landfill. Moreover, asset recovery also is beneficial to cost reduction and product-quality improvement of manufacturing firms, and these benefits can be strengthened if the stakeholders provide supports to the firms' environmental management.

References

1. Carter CRE, Ellram LM (1998) Reverse logistics: are view of the literature and framework for future investigation. J Bus Logist 19(1):85–102
2. Sarkis J (2014) Green supply chain management. Momentum Press, New York
3. Winzelberg D (2012) How reverse logistics makes online returns easy and keeps customers happy. Available from http://www.forbes.com/sites/ups/2012/12/03/how-reverse-logistics-makes-online-returns-easy-and-keeps-customers-happy/
4. Park J, Sarkis J, Wu ZH (2010) Creating integrated business and environmental value within the context of China's circular economy and ecological modernization. J Clean Prod 18 (15):1491–1501
5. Yang JX, Lu B, Xu C (2008) WEEE flow and mitigating measures in China. Waste Manag Res 28(9):1589–1597

6. Flapper SDP, Van Nunen JAEE, Wassenhove LNV (2005) Managing closed-loop supply chains. Springer, New York
7. Rogers D, Tibben-Lembke R (2001) An examination of reverse logistics practices. J Bus Logist 22(2):129–148
8. Stock JR (2001) The 7 deadly sins of reverse logistics. Mater Handl Manag 56(3):5–11
9. Prahinski C, Kocabasoglu C (2006) Empirical research opportunities in reverse supply chains. Omega—Int J Manag Sci 34(6):519–532
10. Lai KH, Wu SJ, Wong CWY (2013) Did reverse logistics hit the triple bottom line of Chinese manufacturers. Int J Prod Econ
11. Heidrich O, Harvey J, Tollin N (2009) Stakeholder analysis for industrial waste management systems. Waste Manag Res 29(2):965–973
12. Sarkis J, Dijkshoorn J (2007) Relationships between solid waste management performance and environmental practice adoption in Welsh small and medium-sized enterprises (SMEs). Int J Prod Res 45(21):4989–5015
13. Souza K (2013) Retail returns, reverse logistics ramp up for the holidays. Available from http://www.thecitywire.com/node/29994#.VFCneTSUeYA
14. Gomez F, Guzman JI, Tilton JE (2007) Copper recycling and scrap availability. Resour Policy 32(4):183–190
15. Thomas VM (2003) Product self-management: evolution in recycling and reuse. Environ Sci Technol 37(23):5297–5302
16. McKinnon A, Browne M, Whiteing A (2012) Improving the Environmental Sustainability of Logistics, 2nd edn. Kogan Page, London, GBR
17. Beamon BM, Fernandes C (2004) Supply-chain network configuration for product recovery. Prod Plan Control 15(3):270–281
18. Sasikumar P, Kannan G, Haq AN (2010) A multi-echelon reverse logistics network design for product recovery—A case of truck tire remanufacturing. Int J Adv Manuf Technol 49 (9):1223–1234
19. Daoud Ait-Kadi MC, Marcotte S, Riopel D (2012) Sustainable reverse logistics network. Wiley, Hoboken, US
20. Bin Zhang ZW, Lai K (forthcoming) Does industrial waste reuse bring dual benefits of economic growth and carbon emission reduction?—Evidence of incorporating the indirect effect of economic growth in China. J Ind Ecol
21. Tudor T, Adam E, Bates M (2007) Drivers and limitations for the successful development and functioning of EIPs (eco-industrial parks): a literature review. Ecolog Econ 61(2):199–207
22. Zhu Q, Sarkis J, Lai K (2014) Supply chain-based barriers for truck-engine remanufacturing in China. Trans Res Part E. 68:103–117
23. Melnyk SA, Sroufe RP, Calantone R (2003) Assessing the impact of environmental management systems on corporate and environmental performance. J Oper Manag 21(3): 329–351
24. Krikke H, Blanc Il, Velde Svd (2004) Product modularity and the design of closed-loop supply chains. Calif Manag Rev 46(2):23–39
25. Kainuma Y, Tawara N (2006) A multiple attribute utility theory approach to lean and green supply chain management. Int J Prod Econ 101(11):99–108
26. Seitz MA, Peattie K (2004) Meeting the closed-loop challenge: The case of remanufacturing. Calif Manag Rev 46(2):74–89
27. Sbihi A, RW Eglese (2007) Combinatorial optimization and green logistics. 4OR 5:99–106
28. Toffel MW (2004) Strategic management of product recovery. Calif Manag Rev 46(2): 120–141
29. Chung C-J, Wee H-M (2008) Green-component life-cycle value on design and reverse manufacturing in semi-closed supply chain. Int J Prod Econ 113(2):528–545
30. Greener Design Staff (2009) Patagonia's clothing recycling program: lessons learned, challenge ahead. Greener Design
31. Miemczyk J (2008) An exploration of institutional constraints on developing end-of-life product recovery capabilities. Int J Prod Econ 115(2):272–282

32. Iyer AV, Bergen ME (1997) Quick response in manufacturer retailer channels. Manag Sci 43 (3):559–570
33. Zhu QH, Sarkis J, Lai K-H (2008) Confirmation of a measurement model for green supply chain management practices implementation. Int J Prod Econ 111(2):261–273
34. Baumann H, Boons F, Bragd A (2002) Mapping the green product development field: engineering, policy, and business perspectives. J Clean Prod 1(4):409–425
35. Savaskan RC, Bhattacharya SH, Van Wassenhove LN (2004) Closed-loop supply chain models with product remanufacturing. Manag Sci 50(2):239–252
36. Savaskan RC, Van Wassenhove LN (2006) Reverse channel design: the case of competing retailers. Manag Sci 52(1):1–14
37. Sroufe R (2003) Effects of environmental management systems on environmental management practices and operations. Prod Oper Manag 12:416–431
38. Wong CWY et al (2012) The roles of stakeholder support and procedure-oriented management on asset recovery. Int J Prod Econ 135(2):584–594
39. Sousa R, Voss CA (2002) Quality management re-visited: a reflective review and agenda for future research. J Oper Manag 20(1):91–109
40. Buzzelli DT (1991) Time to structure an environmental policy strategy. J Bus Strategy 12 (2):17–20
41. Brophy M (ed) (1996) The essential characteristics of an environmental policy. In: Welford R (ed) Corporate environmental management: systems and strategies. Earthscan, London, pp 92–103